Praise for the Clinicians' and Educators' Desk Reference

"I urge my class of fourth-year medical students and residents to refer to the CEDR often in the future. It really is a wonderful, IPE/C-themed guide." – **Shelley R. Adler, PhD**, Professor. Department of Family and Community Medicine; Director of Education, Osher Center for Integrative Medicine University of California, San Francisco

"Congratulations on an excellent, helpful and much needed regulatory road map." – **Richard "Buz" Cooper, MD**, Professor of Medicine and Senior Fellow, Leonard Davis Institute of Health Economics, University of Pennsylvania

"Offers all of us, for the first time, a clear view of the CAM professions: their skills, approaches to patient care, education and accreditation.... extremely useful for current health practitioners who need to know more about each of the CAM disciplines, and for consumers who are evaluating their health-care choices. " – **Elaine Zablocki**, Townsend Letter, November 2010

"This handbook is an excellent resource for all nurses....The book also includes a very informative chapter on Collaboration between the Disciplines." –**American Holistic Nurses Association**, News From AHNA, Vol. 9 No. 3

"The *Clinicians' and Educators' Desk Reference* certainly fills a hole in health education textbook resources....healthcare students of all disciplines need this concise and current summary of this evolving and exciting sector of the healthcare system. It is well-organized, up-to-date and will certainly be used to foster greater interprofessional understanding and inter-disciplinary behavior among healthcare practitioners." – **William Meeker, DC, MPH**, President, Palmer College-West Campus

"This reference text is meant to support students, practitioners, patients and academic and policy leaders in creating appropriate and respectful collaboration between the disciplines to provide optimal healthcare." – **Cherie Sohnen-Moe**, March 2010 Teacher's Aide

"This is a reference book every student interested in CAM should have on their shelf." – **Adam Perlman, MD, MPH**, Past-Chair, Consortium of Academic Health Centers for Integrative Medicine; Associate VP, Duke Health & Wellness; Executive Director, Duke Integrative Medicine

"This reference can be helpful when responding to patient questions regarding other disciplines, as well as when referring patients to those same disciplines. Having additional information about those disciplines might help us to better select which one is a better fit for referral for our patients." – **Marcia Prenguber, ND**, Director of Integrative Care, Indiana University Health Goshen Center for Cancer Care

"In short, the CEDR covers the topics that we need and is the perfect text for our *Survey of Health Professions* course." – **Ed Shaheen, MD,** Chairman of Western Medicine Department, Yo San University of TCM

"Even though I consider myself quite well-informed about this topic, I was impressed with how much I learned from this desk reference. Most of the books about CAM discuss primarily the modalities and not much about the practitioners....Ultimately, therefore, it will give us greater understanding of the CAM professionals which will then support the goal of all practitioners – better health care for our patients." – **Bill Manahan, MD**, Assistant Professor Emeritus, Department of Family Medicine and Community Health, University of Minnesota Academic Health Center

"Congratulations on producing a very useful and much needed reference. Dr. Amri and I decided that all graduates of our Program at Georgetown University in Complementary and Alternative Medicine should receive a copy." – **Aviad Haramati, PhD**, Professor in the Departments of Physiology & Biophysics and Medicine, Georgetown University School of Medicine; Founding Vice Chair, Consortium of Academic Health Centers for Integrative Medicine

"This well-referenced book will appeal to those clinicians with a scientific- and evidence-based mindset, and yet is written in simple enough language so that no therapist, or client, would be intimidated by it.... I think it's important to get this book into the hands of as many mainstream-medicine practitioners as possible." – **Laura Allen**, In Massage Today, February 2010

"Consolidates hard to find information on specific issues for each of the disciplines in one go-to place." – **Ruth Levy, MBA, CP AOBTA**, Director of Admissions, Pacific College of Oriental Medicine

"The desk reference is intended for clinicians interested in collaboration and referral and will also serve as a textbook and useful resource for educators and students." –**Midwifery Education Accreditation Council**, MANA NEWS, November 2009

"We are giving these books to our full time and pro-rata faculty at our Faculty Development Day. Faculty need to have access to this text before they can envision how they might use it. Once faculty have had a chance to review the book, it is likely they will find innovative ways to incorporate it in their courses and share it with other providers." – **Renée M. DeVries, DC, DACBR**, Dean, College of Chiropractic, Northwestern Health Sciences University

"A guide to the five licensed CAM professions, written by vetted experts, that can be used as a desk reference for clinicians and a textbook and resource for educators and students." – **Integrative Practitioner Newsletter,** March 2010

"This essential desk reference correctly states that it is time for traditional health care to become more integrated with complementary therapies, and that one of the primary blocks to that has been the lack of knowledge of traditional providers about the licensed professions that provide alternative care." – **Michael H. Cohen, Esq.**, Former Assistant Professor of Medicine, Harvard Medical School; Publisher, CAM Law Blog

"The CEDR and companion power point presentation are fantastic resources - my thanks to ACCAHC for producing them; I refer my conventional colleagues to the CEDR quite regularly." – **Erica Oberg, ND, MPH**, Director of Clinical Services at Bastyr University Clinic in San Diego, Director of the Bastyr University Center for Health Policy and Leadership

"This is a useful text that will be employed frequently in the day-to-day care of patients, and will certainly remain a valued reference for each of our students long after graduation." – **Joseph E. Brimhall, DC,** President, University of Western States

"A concise, well-articulated and reliable reference that serves as a landmark step in fostering collaboration and understanding between healthcare professions.... This excellent desk reference will be valuable for healthcare professionals, institutions, schools, educators, students, libraries and patients. " – **Lori Howell, LAc, DAOM**, Former Director of Clinical Services, Pacific College of Oriental Medicine; Former Assistant Academic Dean, Pacific College of Oriental Medicine

Clinicians' and Educators' Desk Reference on the Licensed Complementary and Alternative Healthcare Professions

Second Edition

Developed by
Academic Consortium for Complementary and Alternative Health Care

Partner Organizations

Alliance for Massage Therapy Education
Association of Accredited Naturopathic Medical Colleges
Association of Chiropractic Colleges
Council of Colleges of Acupuncture and Oriental Medicine
Midwifery Education Accreditation Council

Project Managers and Editors

Elizabeth Goldblatt, PhD, MPA/HA, ACCAHC Chair
Pamela Snider, ND, ACCAHC Founding Executive Director
Beth Rosenthal, MPH, MBA, PhD, ACCAHC Assistant Director
Sheila Quinn, Consulting Editor
John Weeks, ACCAHC Executive Director

www.accahc.org

Acknowledgements

We gratefully acknowledge our partner organizations and the authors and editors of these chapters. These are noted in the Table of Contents and with their relevant sections. Three others provided welcome input that is not otherwise noted: Kate Henrioulle Zulaski and James Winterstein, DC, offered helpful insights on review of chapters, and Renée Motheral Clugston provided expert assistance in producing the book. Finally, the book's original design, layout, printing, marketing and distribution were made possible by very generous grants from National University of Health Sciences and Lucy Gonda. This book took a community of collaborators.

Clinicians' and Educators' Desk Reference on the Licensed Complementary and Alternative Healthcare Professions

©2013, Academic Consortium for Complementary and Alternative Health Care (ACCAHC)
3345 59th Avenue SW
Seattle, Washington 98116
www.accahc.org

Disclaimer:
This publication is intended for use as an educational tool only and should not be used to make clinical decisions concerning patient care. Information is subject to change as educational and regulatory environments evolve. In addition, the role of the partner organizations was to identify authors. These organizations have not endorsed the content of the book, and their participation in selecting the chapter authors was not intended as an endorsement.

ISBN 978-1-304-08294-7

Table of Contents

Continued on next page

Appendices:

Preface

Chiropractic, massage, acupuncture and Oriental medicine, midwifery, and naturopathic medicine were in common use before national surveys documented their sizeable footprint within the US healthcare delivery system.

Four national surveys conducted since 1990 have found that at least a third of US adults routinely use complementary and alternative medical (CAM) therapies to treat their principal medical conditions. All four surveys documented the fact that each year Americans schedule hundreds of millions of office visits to licensed complementary and alternative healthcare professionals at a cost of tens of billions of dollars, most of which is paid for out-of-pocket. The most recent national survey published by the Centers for Disease Control (NHIS 2007) reported that out-of-pocket expenditures associated with complementary and alternative healthcare practices account for approximately 12% of all out-of-pocket health care expenditures in the United States. The discovery of this "hidden mainstream" of American health care is no longer disputed.

These national surveys have also confirmed that the majority of patients who use complementary and alternative medical therapies generally do not disclose or discuss them with their primary care physicians or subspecialists. This lack of interdisciplinary communication is not in a patient's best interest. Making matters more complicated, it is also known that among individuals who use CAM therapies, the majority tend to use multiple CAM practices as opposed to a single modality to treat their medical problems. Regrettably, many individuals seek combinations of conventional and complementary therapies in a non-coordinated rather than an integrated fashion.

The Institute of Medicine, in its report *Complementary and Alternative Medicine in the United States*, has helped set the stage for future research in this area. The report states:

> Studies show that patients frequently do not limit themselves to a single modality of care – they do not see complementary and alternative medicine (CAM) and conventional medicine as being mutually exclusive – and this pattern will probably continue and may even expand as evidence of therapies' effectiveness accumulate. Therefore, it is important to understand how CAM and conventional medical treatments (and providers) interact with each other and to study models of how the two kinds of treatments can be provided in coordinated ways. In that spirit, there is an urgent need for health systems research that focuses on identifying the elements of these integrative medical models, their outcomes and whether these are cost effective when compared to conventional practice.
>
> —Institute of Medicine
> *Report on Complementary and Alternative Medicine in the US*
> National Academy of Sciences, 2005

In this era of cost containment and healthcare reform, in addition to the development of patient centered medical homes and accountable care organizations, this recommendation needs to become a priority. In order to design and implement such studies, however, we must first define and describe each of the relevant CAM modalities and professional groups in order to establish standards that lend themselves to reproducible research. Moreover, referrals between professional communities and the training of multidisciplinary teams consisting of both conventional and complementary care practitioners cannot begin unless and until all participating professional groups learn more about òne another.

This *Clinicians' and Educators' Desk Reference on the Licensed Complementary and Alternative Healthcare Professions* contributes enormously to the goal of designing, testing, and refining models of multidisciplinary, comprehensive, integrative care. It thoughtfully yet concisely describes each of the major complementary and alternative healthcare professions. This information will be useful to patients, healthcare professionals, educators, students, and those responsible for future clinical research and healthcare policy. The Academic Consortium for Complementary and Alternative Health Care is to be commended for making this information readily available.

There is a Chinese proverb which reads, "The methods used by one will be faulty. The methods used by many will be better." Over the next decade, let us test this idea, collaboratively and with all the scientific rigor, clinical expertise, and integrity we have to offer. The next generation will surely thank us.

David Eisenberg, MD
Associate Professor, Harvard School of Public Health
Executive Vice President for Health Education and Research, Samueli Institute
Former Chair, Division of Research and Education in Complementary and Integrative Medical Therapies, Harvard Medical School

Foreword

This handbook is arriving on the healthcare scene in the midst of tumultuous debate about reforms in the US health system. As such, it contributes to the dialogue in a substantive way by providing, logically and clearly, well-organized descriptions of the major complementary healthcare professions.

Three issues received considerable attention in a recent Institute of Medicine Summit on Integrative Medicine and the Health of the Public (February 2009):

- The US has an inadequate primary care workforce. Increasingly, there is recognition that there are licensed CAM practitioners who are prepared to be the first point of entry for many patients. There are also advanced practice nurses who can serve as primary care providers. Current workforce planning needs to take into account a broader range of providers.
- There was also considerable discussion regarding the need to re-orient the system from one that focuses on disease to one that also promotes health. CAM providers have much to contribute in both health promotion and disease treatment.
- Given the complexity of disease and the multiple challenges to human health, rarely can one provider meet all of the physical, emotional, psychological, and spiritual needs of a patient. A team approach that takes into account the capacities of a full range of providers, including CAM practitioners, is critical.

We believe that the time is right for a well-planned and deliberate strategy to integrate CAM practitioners into the nation's healthcare system. These providers are well positioned to focus on prevention, lifestyle change, health coaching, and management of certain chronic diseases and painful conditions. This type of integration has the potential to increase quality while decreasing costs of health care.

This handbook offers an orientation for the healthcare teams of the future. Many conventional providers simply do not know enough about the training, scope of practice, and evidence supporting many of the complementary therapies and practitioners described here. This book helps to bridge that information gap.

In the future, we would like to see more practical and detailed descriptions of interdisciplinary team collaborations between physicians, advanced practice nurses, physician assistants, and physical therapists from the conventional medical world and the CAM practitioners listed in this handbook. The processes facilitating referral, interdisciplinary communication, and team building are perhaps the topic of a follow-up handbook that will be essential for the next phase of evolution into true interdisciplinary and transdisciplinary healthcare teams.

As members of the Consortium of Academic Health Centers for Integrative Medicine, our work has involved beginning the process of bringing mainstream medicine, nursing, and allied health professionals into a positive relationship with complementary therapies and disciplines. Much remains to be done in these areas.

However, the benefit of a handbook such as this one is that it creates a common ground in which patient-centered care is central and in which respect for others with professional skills and perspectives different from our own is encouraged. We see this type of publication as a helpful textbook and reference not only to the CAM community but to the conventional healthcare professions as well.

Victor S. Sierpina, MD, ABFM, ABIHM
Past Chair, Consortium of Academic Health Centers for Integrative Medicine
Director, Medical Student Education
WD and Laura Nell Nicholson Family Professor of Integrative Medicine
Professor, Family Medicine
University of Texas Distinguished Teaching Professor
University of Texas Medical Branch

Mary Jo Kreitzer, PhD, RN, FAAN
Past Vice-Chair, Consortium of Academic Health Centers for Integrative Medicine
Director, Center for Spirituality and Healing
Professor of Nursing
University of Minnesota

Adam Perlman, MD, MPH, FACP
Past Chair, Consortium of Academic Health Centers for Integrative Medicine
Associate VP, Duke Health & Wellness
Executive Director, Duke Integrative Medicine

December 2012

Introduction: Deepening Engagement in Interprofessional Practice and Education

John Weeks and Elizabeth Goldblatt, PhD, MPA/HA

The Institute of Medicine (IOM) of the National Academy of Sciences published *Complementary and Alternative Medicine in the United States* in 2005. The authors described the context of emerging integration behind the IOM exploration and the "overarching rubric" that should guide the integration of these practices with conventional medicine:

> The level of integration of conventional medicine and complementary and alternative medicine (CAM) therapies is growing. That growth generated the need for tools or frameworks to make decisions about which therapies should be provided or recommended, about which CAM providers to whom conventional medical providers might refer patients, and the organizational structure to be used for the delivery of integrated care. The committee believes that the overarching rubric that should be used to guide the development of these tools should be the goal of providing comprehensive care that is safe and effective, that is collaborative and interdisciplinary, and that respects and joins effective interventions from all sources.

> —Institute of Medicine
> *Complementary and Alternative Medicine in the United States*
> National Academy of Sciences, 2005, p. 216

We offered the 2009 edition of this book in the spirit of the IOM rubric expressed here. This book is also intended for use by well-educated consumers of health care who are frequently building their own teams of clinicians from a variety of health professions.

We publish this second edition to keep the information current, and in the context of the exciting, hopeful, and relatively recent movement for interprofessional practice and education that has arisen since. The vision and mission of the Academic Consortium for Complementary and Alternative Health Care (ACCAHC) since the founding of the organization as a multidisciplinary initiative has articulated our commitment to a healthcare system based in appropriate team care.

Vision

ACCAHC envisions a healthcare system that is multidisciplinary and enhances competence, mutual respect and collaboration across all healthcare disciplines. This system will deliver effective care that is patient centered, focused on health creation and healing, and readily accessible to all populations.

Mission

The mission of ACCAHC is to enhance the health of individuals and communities by creating and sustaining a network of global educational organizations and agencies, which will promote mutual understanding, collaborative activities and interdisciplinary healthcare education.

ACCAHC is pleased to be involved with multiple national and global projects related to this promotion of teamwork for the benefit of the patient and the public. These include: development of appropriate competencies for practice, education and research; sponsorship of, and participation in, the new Institute of Medicine Global Forum on Innovation in Health Professional Education; involvement in the Collaboration Across Borders conferences; a steering committee role and active engagement in a national campaign to promote an integrated approach to pain treatment; and most recently, participating in the new Center for Interprofessional Practice and Education.

Our work focuses on the five licensed complementary and alternative healthcare professions, the members of which are providing the lion's share of integration of complementary and alternative medicine (CAM) among community providers, in clinics and hospitals, insurance and healthcare benefit designs, research and educational grants, government programs, and academic centers throughout the United States. The five are acupuncture and Oriental medicine, chiropractic, massage therapy, direct-entry midwifery, and naturopathic medicine. These professions have all achieved an advanced level of regulatory maturity, including attaining the important professional benchmark of earning United States Department (Secretary) of Education recognition for the agencies that accredit their educational institutions. These five complementary healthcare disciplines represent roughly 375,000 practitioners nationwide.

In order to support expanded understanding, informed referrals, and choices across the entire integrative practice domain, we also offer information on the related fields of yoga therapy, holistic medicine, holistic nursing, integrative medicine, homeopathy, and Ayurvedic medicine. These related integrative practice fields were chosen based on their participation in the 2005 National Education Dialogue to Advance Integrated Health Care.

Together, the practitioners from the licensed CAM professions and related fields represent the integrative healthcare professionals whom a practitioner is most likely to encounter in developing a more integrated team of community resources for patients. They are also the integrative practice professionals who may be seeing the patients of many conventional providers, with or without their knowledge.

We are pleased to witness the excitement and acclaim with which this book was received by many educators and reviewers across multiple disciplines. At the same time, we were surprised that the book's use in program curricula has thus far, while growing, been limited. In an effort to increase use of the book in program curricula, we are making sample curriculum available to academic programs on our website. To make sure that neither format nor price is an obstacle to use of the book, we are making this edition of the CEDR freely available in a downloadable PDF format. Hard copies are available for purchase for those who wish them. The book also includes a companion slide show for educators that is available via the ACCAHC website

We hope that this book, in its new form and in our future editions, continues to be a useful "tool and framework", in the words of the IOM text, not only to "make decisions about CAM providers to whom conventional medical providers might refer patients", but also to ensure that members of the distinct disciplines whose practices are portrayed here are well informed about each other.

Supporting Multidisciplinary Care: A Focus on Practitioners Rather than Modalities

In this era in which the need to focus on wellness, health promotion and disease prevention for better treatment of chronic disease dominates the healthcare reform dialogue, multidisciplinary teams form the foundation of most of the significant new clinical models. A team approach to health care provides higher patient satisfaction and better health care, and reduces costs. Advocates of the chronic care model, of prospective health care, of the medical home, of integrative medicine, and of functional medicine, all promote an increasingly multidisciplinary medicine. The movements for interprofessional practice and education and integrated pain care are similarly focused. Success in such teams begins with quality information about the skills and background of potential contributors from diverse professions.

Practitioners of all types tend to be educated in silos. We have a saying in the Academic Consortium for Complementary and Alternative Health Care (ACCAHC) that "those who are educated together practice together." If we wish to develop a patient-centered healthcare system, then we must break out of these isolated silos and learn more about one another.

This book is distinguished by its focus on practitioners and their philosophies. Our authors explore approaches to patients, education, accreditation, scopes of practice, and professional developments. We worked with key national educational organizations to select authors whose rich experience as clinicians and educators would inform the writing. We believe that in this day and age when patients are putting together their own healthcare teams, that all healthcare professionals have the ethical responsibility to become more educated about the licensed CAM fields.

This orientation is in contra-distinction to most education and publications relative to complementary and alternative practices. Past contributions relative to "complementary and alternative medicine" (CAM) focused mainly on individual agents, modalities, or therapies. Insufficient attention has been given to exploration of the whole practices of these disciplines and the systems of care that the distinctly licensed CAM practitioners are educated to provide. Too often, the portraits of these professions, when they are offered, come from outside observers rather than from those with personal experience as practitioners.

The historically limited focus of most texts targeting multidisciplinary audiences does a disservice to both practitioners and consumers. A description of a single natural therapy, for instance, will not teach about the context in which it is offered by an actual practitioner. The reader will not encounter the profession's distinct approach to patients, the depth of the professional education which informs it, and the sometimes surprising scopes of practice that underlie it. Approaches may include conventional and nonconventional diagnostic strategies, together with diverse treatments from their own disciplines.

In addition, a reductive focus on CAM therapies for specific diseases, such as a botanical or the use of acupuncture needles, does not teach a community practitioner that the healthcare disciplines portrayed in this book tend also to emphasize preventive medicine, the practitioner-patient relationship, the promotion of wellness, and the teaching of self-care practices. As became abundantly clear through the ACCAHC project on these disciplines and primary care, these practitioners tend also to be "first choice" or "first access" providers for the patients who come to these. These disciplines reflect whole systems of care rather than more limited modalities or therapies. The loss of knowledge and skills in using a reductive focus can be significant. An excellent perspective on potential problems is provided by the IOM report on CAM cited above:

> Although some conventional medical practices may seek and achieve a genuine integration with various CAM therapies, the hazard of integration is that certain CAM therapies may be delivered within the context of a conventional medical practice in ways that dissociate CAM modalities from the epistemological framework that guides the tailoring of the CAM practice. If this occurs, the healing process is likely to be less effective or even ineffective, undermining both the CAM therapy and the conventional biomedical practice. This is especially the case when the impact or change intended by the CAM

therapy relies on a notion of efficacy that is not readily measurable by current scientific means. (p. 175)

A modality or therapy orientation also falls short in assisting students, educators, and clinicians to understand how to communicate with and work in teams with other practitioners, and renders a disservice to consumers as well. Patients are too often treated by professionals who have neither the knowledge nor comfort to work well with members of these complementary healthcare professions who may offer what will serve the patient best. This book, while foundational, is also therefore remedial.

Additional Context for this Book

We have taken a perspective that is part of a broader shift in the integrative practice and CAM dialogue, just as it is part of the heightened focus on interprofessional practice and education in the broader health reform effort. In 2005, for instance, the Consortium of Academic Health Centers for Integrative Medicine (CAHCIM), consisting now of 54 member institutions, changed its operating definition of integrative medicine. The amended definition emphasizes the importance of integrating not only "modalities" but also "healthcare professionals and disciplines." (See the Definition of Integrative Medicine at www.imconsortium.org/about/home.htm.) Notably, the interprofessional education movement stresses that "egalitarianism" is a core characteristic of the model that infuses optimal team care. All disciplines may not contribute equally to a patient's care but the relationship between them is one of respect for what each brings to the patient. This is a core ACCAHC value.

The commitment of educators and accreditation officials from the five licensed disciplines featured in this book to form ACCAHC as an independent organization is also evidence of the emerging multidisciplinary universe. While the artifact called

"CAM" neatly bundled these distinct disciplines into a group, in truth, these professions also emerged in separate silos and each offers a distinct approach to health care. ACCAHC represents the growing recognition of leaders in these fields that their students are increasingly entering an era that values knowledge and skills in multidisciplinary teams. Membership in ACCAHC reflects a commitment to developing new knowledge, skills, and attitudes that promote collaboration and patient-centered health care.

The research environment also offers evidence of this shift toward broader dialogue from the limited focus on the use of individual agents. Attendees of the 2009 North American Research Conference on Complementary and Integrative Medicine, the most significant research meeting for integrative practice, witnessed a noticeable increase in program content relative to whole systems of care, patient-centered outcomes in care that focus on the whole person, and the challenges in capturing outcomes of actual practices. (See www.cahcimabstracts.com.) The 2011-2015 Strategic Plan of the National Institutes of Health National Center for Complementary and Alternative Medicine (NCCAM), marks a shift toward effectiveness research and real world outcomes. The ACCAHC-NCCAM dialogue successfully supported elevation of the importance of researching the impact of healthcare disciplines on populations. (See ACCAHC Research Working Group Led Communication with NIH NCCAM on the 2011-2015 Strategic Plan: (http://accahc.org/images/stories/2011-15_strat_plan.pdf)

This refocusing of the integration dialogue is also reflected in the Patient Protection and Affordable Care Act (2010). That law directly includes "licensed complementary and alternative medicine practitioners", and "integrative health practitioners" and "integrative health" in sections related to workforce, medical homes, prevention and health promotion, patient-centered outcomes research and, indirectly, in a requirement for non-discrimination among provider types in health plans. The language establishing a new focus on wellness and prevention is aligned with the health orientation and wellness principles of the licensed CAM professions and related integrative practice disciplines.

We believe that the time is ripe for this practical desk reference. We have organized this book to facilitate ready access to useful information. We continue in this introduction with some general information on the five licensed complementary healthcare professions. This leads directly into the chapters from each of these five licensed integrative healthcare professions. Similar information on the six related integrative practice fields is found in later chapters. We share additional information about ACCAHC in appendices. We include some accomplishments since ACCAHC was founded as a project of the Integrative Healthcare Policy Consortium (www.ihpc.org) in 2004. Given ACCAHC's organizational structure and focus as an interprofessional practice, we provide a look at some of the collaborative, interprofessional activity with which we are proud to be involved.

Data on the Licensed Complementary Healthcare Professions

As noted above, the five licensed complementary healthcare professions have each distinguished themselves by achieving certain benchmarks of professional development. These benchmarks include creating councils of colleges or schools, finding agreement on consistent national educational standards, establishing an accrediting agency, and then earning recognition of the accrediting agency by the US Department (Secretary) of Education. Each achievement marks a stage of maturation.

In addition, each of these licensed professions has also created a national certifying or testing agency, which is important to the public and to state regulators as the profession begins the long process of promoting licensing in each state. To expand licensing, initial groups of practitioners in each state formed state associations, mustered political skills and courage, and created coalitions of supportive patients and consumers to mobilize members of one legislature after another to enact licensing statutes for these professions.

Table I.1

Development of Standards by the Licensed Complementary and Alternative Healthcare Professions

Profession	Accrediting agency established	US Department of Education recognition	Recognized schools or programs	Standardized national exam created	State regulation*	Estimated number of licensed practitioners in the US
Acupuncture and Oriental medicine	1982	1990	61	1985	44 states + DC	28,000
Chiropractic^	1971	1974	15	1963	50 states + DC, Puerto Rico, and all other US territories /insular areas	72,000
Massage therapy	1982	2002	88	1994	44 states + DC	280,000
Direct-entry Midwifery	1991	2001	10	1994	26 states	2,000
Naturopathic medicine	1978	1987	7	1986	16 states, DC, Puerto Rico, Virgin Islands	5,500

*For chiropractors and naturopathic physicians, this category uniformly represents licensing statutes; for acupuncture, virtually all states use licensure; for massage, there is a mixture of licensing, certification, and registration statutes.
^Although the Council on Chiropractic Education was first incorporated in 1971, there were other accrediting agencies and activities within the chiropractic profession dating back to 1935.

Success in achieving these benchmarks is a prerequisite for integration into the broader healthcare enterprise in the US. Members of these disciplines are able to meet core credentialing standards for participation in conventional payment and delivery systems. Table 1.1 provides data on the status of each profession in these benchmarking processes. As is clear, while all five disciplines have a federally-recognized accrediting agency, the number of states in which licensing has been established varies significantly.

Partnerships with National Organizations for Content

Many questions arise when considering a publication to describe a number of disparate professions. How can one most equitably develop a respectful and useful resource? Who should be charged with authoring the chapters? How can we ensure balance and accuracy?

We chose to address these questions by working with and through authors selected by leading national educational organizations affiliated with the five professions. For the core chapters we contacted the councils of colleges and schools that are organizational members of ACCAHC. For direct-entry midwifery, we worked with that profession's accrediting agency. While these organizations helped us identify authors with whom we worked to completion of the book, the partner organizations did not endorse the content of this section, their chapters, or the book as a whole.

For the related integrative practice fields, we secured authors who were selected by the leading professional organization in the field. For integrative medicine, which is not yet a profession per se, we worked with the Consortium of Academic Health Centers for Integrative Medicine. Table 1.2 lists the chapter partner organizations, authors and editors. We are deeply grateful for all their contributions.

Table I.2

Partner Organizations and Authors

Chapters	Partner Organizations	Authors & Editors
Acupuncture and Oriental Medicine	Council of Colleges of Acupuncture and Oriental Medicine	David Sale, JD, LLM Steve Given, DAOM, LAc Catherine Niemiec, JD, LAc Elizabeth Goldblatt, PhD, MPA/HA Kory Ward-Cook, PhD, MT (ASCP), CAE** Vitaly Napadow, PhD, LAc**
Chiropractic	Association of Chiropractic Colleges	Reed Phillips, DC, PhD* Michael Wiles, DC, MEd, MS David O'Bryon, JD, CAE Joseph Brimhall, DC**
Massage Therapy	American Massage Therapy Association Council of Schools (2009 edition) Alliance for Massage Therapy Education (2013 edition)	Jan Schwartz, MA Cherie Monterastelli, RN, MS, LMT* Stan Dawson, DC, NMT, CMT** Pete Whitridge, BA, LMT**
Direct-entry Midwifery	Midwifery Education Accreditation Council	Jo Anne Myers-Ciecko, MPH
Naturopathic Medicine	Association of Accredited Naturopathic Medical Colleges	Paul Mittman, ND, EdD Patricia Wolfe, ND* Michael Traub, ND, DHANP, FABNO Christa Louise, MS, PhD** Marcia Prenguber, ND** Elizabeth Pimentel, ND, LMT**
Related integrative practice fields		
Ayurvedic Medicine	National Ayurvedic Medical Association	Felicia Tomasko, RN
Holistic Medicine	American Holistic Medical Association	Hal Blatman, MD* Kjersten Gmeiner, MD* Donna Nowak, CH, CRT* Molly Roberts, MD, MS** Steve Cadwell**
Holistic Nursing	American Holistic Nurses Association	Carla Mariano, EdD, RN, AHN-BC, FAAIM
Homeopathy	American Medical College of Homeopathy (2009 edition) Accreditation Commission for Homeopathic Education in North America (2013 edition)	Todd Rowe, MD* Heidi Schor**
Integrative Medicine	Consortium of Academic Health Centers for Integrative Medicine	Victor Sierpina, MD* Benjamin Kligler, MD, MPH**
Yoga Therapy	International Association of Yoga Therapists	John Kepner, MBA

* Author for 1st edition only   ** Editor for 2nd edition only

Templates for the Chapters

The author or author teams for the core chapters agreed to write approximately 6000 words according to a template that was intended to make this an easy reference guide for readers. For the related integrative practice fields, we used a similar template, but asked for 1000 words. We gave the authors flexibility in the extent to which each would choose to focus on an individual section from among the following:

Philosophy, Mission, Goals
Characteristics and Data
Clinical Care
    Approach to patient care
    Scope of practice
    Referral practices
    Third-party payers
Integration activities
Education
    Schools and programs
    Curriculum content
    Faculty and other training information
    Accreditation
Regulation and Certification
    Regulatory status
    Examinations and certifications
Research
Challenges and Opportunities
Resources
    Organizations and websites
    Bibliography
    Citations

The authors understood that their chapters would have to pass muster beyond the borders of their own profession. The content was reviewed by an internal editing team and also by the authors from the other disciplines. If any reviewer felt that another discipline had overstated its value, for instance, the chapter authors were asked to make edits or make their case stronger. Numerous changes were made during this review process.

Collaboration: The Birthing of this Book and the Hope for Its Future

We offer this book as a tool to facilitate greater understanding and collaboration between well-informed practitioners, educators and students, and to support better health care for patients. Respectful collaboration has been our vision and mission since ACCAHC was first conceived. The Appendix of this book shares some advances in cross-disciplinary education and practice over the past half-decade in which ACCAHC has been proud to be involved.

This book also represents collaboration with donors and with organizations that enhance our ability to fulfill our mission.

We are deeply grateful to Lombard, Illinois-based National University of Health Sciences (NUHS-www.nuhs.edu) for a significant project grant which allowed us to bring this multi-year project to completion in the 2009 edition. Specifically, we thank James Winterstein, DC, president of NUHS for his support for this grant. NUHS has programs in four of the licensed professions discussed here: chiropractic medicine, acupuncture and Oriental medicine, naturopathic medicine, and massage therapy. ACCAHC also thanks the Council of Colleges of Acupuncture and Oriental Medicine (www.ccaom.org) for a 2007 contribution that supported initial work on this project.

Critical financial support which nurtured ACCAHC through this gestational process has come from Lucy Gonda, who is, in essence, an ACCAHC co-founder. Her vision and philanthropic support gave us our first boost into life in 2003. She continues to provide key support for our mission including significant contributions in 2011 that allowed us to place the first edition of this book in the hands of leading educators in interprofessional education across North America. Joining Gonda in the ACCAHC Sustaining Donors Group which supported relationship development and staff time for this book are the Westreich Foundation, the Leo S. Guthman Fund, NCMIC Foundation, Bastyr University and Life University. We thank, in particular, Ruth Westreich, Lynne Rosenthal, Lou Sportelli, DC, Daniel Church, PhD, and Guy Riekeman, DC respectively, for their decisions to move these organizations to back this work.

Finally, this book is now in its second edition. We anticipate that our collaborators from the councils of colleges and schools and the professional organizations will continue to have ideas about how to enhance their chapters once they see the book in print. We would like to hear from you about how we can make this book and accompanying materials more useful. Ideas can be sent to info@accahc.org.

The publication of the second edition of this book comes amidst a new era of interprofessional respect and practice. More individuals, educators and health systems are realizing that optimal team care means widening the circle of involved professionals. Thanks to each of you for your own commitment to expanding your educational or clinical horizons to better understand and respect members of these disciplines. Patients will thank us all for it.

Section I
Licensed Complementary and
Alternative Healthcare Professions

Acupuncture and Oriental Medicine

Chiropractic

Massage Therapy

Direct-Entry Midwifery

Naturopathic Medicine

Acupuncture and Oriental Medicine

First Edition (2009) Authors:
David Sale, JD, LLM; Steve Given, DAOM, LAc;
Catherine Niemiec, JD, LAc; Elizabeth Goldblatt, PhD, MPA/HA

Second Edition (2013) Editors:
Kory Ward-Cook, PhD, MT (ASCP), CAE; Elizabeth Goldblatt,
PhD, MPA/HA; David Sale, JD, LLM; Vitaly Napadow, PhD, LAc;
Catherine Niemiec, JD, LAc; Steve Given, DAOM, LAc

Partner Organization: Council of Colleges of Acupuncture and
Oriental Medicine (CCAOM)

About the Authors/Editors: Sale is Executive Director of the Council of Colleges of Acupuncture and Oriental Medicine. Given is the Dean of Clinical Education at the American College of Traditional Chinese Medicine and a past member of the Accreditation Commission for Acupuncture and Oriental Medicine. Niemiec is Chair of the Accreditation Commission for Acupuncture and Oriental Medicine and President of the Phoenix Institute of Herbal Medicine and Acupuncture. Goldblatt is Past President of Council of Colleges of Acupuncture and Oriental Medicine, Doctorate of Acupuncture and Oriental Medicine Council and Faculty member at the American College of Traditional Chinese Medicine, and Chair of Academic Consortium of Complementary and Alternative Health Care. Ward-Cook is the Chief Executive Officer for the National Certification Commission for Acupuncture and Oriental Medicine. Napadow is Co-President of the Society for Acupuncture Research and Assistant Professor at Harvard Medical School and the Department of Radiology at Massachusetts General Hospital.

Philosophy, Mission, Goals

The history of acupuncture and Oriental medicine (AOM) extends back over 3000 years, with documentation of accumulated knowledge and experience appearing before the Han Dynasty (206 BCE to 220 CE), thus providing a long record of traditional use. Since the 1970s, AOM has experienced increasing popularity in the US.

The philosophy of this ancient therapeutic system is based on the concept of *qi* 气 (simplified) or 氣 (traditional), pronounced "chee", meaning energy/life force, and its flow through the body along channels or meridians. This ancient medicine, which can interface well with conventional medicine, has its own nomenclature, physiology, pathology, and therapeutics, which create a complex system of medicine documented in classical and modern texts.

Practitioners, educators, and national AOM organizations seek to promote a variety of complementary goals. These include: standards of excellence, competence, and integrity in the practice of the profession; public safety; excellence in AOM education; high quality health care; integration of AOM into the US healthcare system; chronic disease amelioration; a preventive health measure, especially as it impacts public health involvement; and assistance to communities affected by disaster, war, conflict, and poverty. Another goal of the AOM community of leaders is to provide evidence as to the efficacy and safety of acupuncture, Chinese herbology and Asian bodywork therapy through research and education.

Characteristics and Data

Prior to the 1970s, AOM was practiced in the US within Asian communities that arose after the completion of the Transcontinental Railroad in the 1840s. Acupuncture experienced a renaissance in the US in the early 1970s. Interest was heightened when James Reston, a newspaper columnist for the *New York Times*, wrote an article about the benefits of acupuncture he received during his recovery from an appendectomy in China when former President Nixon made his historic visit to China in 1971[1]. The following four major acupuncture institutions were formed in the early 1980s:

- Accreditation Commission for Acupuncture and Oriental Medicine (ACAOM), the US Department of Education-recognized national accrediting organization for AOM schools
- American Association of Acupuncture and Oriental Medicine (AAAOM), the national professional association of AOM Practitioners
- Council of Colleges of Acupuncture and Oriental Medicine (CCAOM), the voluntary national membership association for AOM schools
- National Certification Commission for Acupuncture and Oriental Medicine (NCCAOM), the national testing and certification organization for certifying competency of AOM practitioners.

There are approximately 27,000 to 29,000 licensed AOM practitioners throughout the US. The average (gross) income earned by licensed acupuncturists is $60,000; however, it is not uncommon for practitioners to earn in excess of this amount, with reported salaries in some instances exceeding $100,000.[2]

In 2011, there were approximately 8,000 students enrolled in AOM programs in the US.[3] In the early years of the profession in the US, students were primarily those looking for a second career. Increasingly, students today are looking to the AOM field as a first career. Although many of the graduates entering the AOM profession go into solo practice, there are increasing numbers in integrated healthcare delivery organizations, such as outpatient clinics, hospitals, and community clinics.

Clinical Care

Approach to patient care
Treatments provided by AOM practitioners identify a pattern or multiple patterns of disharmony within a patient and redress that disharmony in a variety of ways that may include any or all of the tools of Oriental medicine. AOM seeks to achieve balance by restoration of the harmonious movement of *qi* through the application of opposite energetic forces (e.g., clear heat through administration of cooling herbal formulas or acupuncture points specifically noted for their ability to release heat from the body) or through the strengthening of weak or deficient areas. The AOM practitioner typically proceeds by observing signs and symptoms that comprise a pattern of disharmony and one's constitutional state, rather than by developing a biomedical diagnosis based on etiologic assessments. AOM is used for chronic disease, prevention and wellness, and acute care. Treatments provided by AOM practitioners are appropriate for in-depth individual care as well as for simultaneous treatment of large groups of people, such as treatment of emergency care workers at the site of the 9/11disaster and Hurricane Katrina, drug users in addiction recovery groups, and veterans suffering from PTSD. In addition, acupuncture is now utilized on the battlefield for pain control and a recent movement has seen many practitioners focus on delivering services in community treatment rooms.

Practitioners of AOM approach patients from a holistic perspective. They take into account all aspects of their health, recognizing the interconnectedness of the mind, emotions, and body, as well as the environment. When *qi* is stagnant or out of balance, which can result from natural, physical or emotional causes, illness and premature aging can occur. Oriental medicine involves a variety of techniques to restore balance to *qi* that has become stagnant or imbalanced, including acupuncture and related therapies (various needle techniques and tools, cupping, *gwa sha* (or scraping technique), moxibustion, modern use of electrical stimulation and cold laser therapy; Asian forms of massage such as acupressure, shiatsu, and tui na; herbal medicine; exercise (*tai qi* and *qi gong*); Oriental dietary therapy; and meditation. Unique methods of diagnosis include the ability to interpret topography of the tongue, palpation of pulse and acupuncture points, and extensive observation and interpretation of all bodily symptoms and physical features according to Oriental medical physiology and diagnostic criteria. This medicine is used to treat a broad range of illnesses ranging from rehabilitation and pain management to virally mediated disorders and neurological complaints. There are several styles of Oriental medicine, including Traditional Chinese Medicine (TCM), Worsley/5-element theory, Japanese meridian theory, French Energetics, and Korean hand therapy.

Practitioners generally spend a significant amount of time developing a collaborative relationship with their patients, assisting them in maintaining their health and promoting a consciousness of wellness. AOM may be especially appropriate for reducing healthcare costs and improving access to care.[4] It is also very safe, resulting in very rare unintended events.[5]

Scope of practice

The scope of practice for AOM varies from state to state. In state statutes acupuncture is frequently defined as *the stimulation of certain points on or near the surface of the human body by the insertion of needles to prevent or modify the perception of pain, to normalize physiological functions, or to treat certain diseases or dysfunctions of the body.* A number of state statutes also reference the *energetic* aspect of acupuncture by noting its usefulness in controlling and regulating the flow and balance of *qi* in the body or in normalizing energetic physiological function. Other state statutes may define acupuncture by reference to traditional or modern Chinese or Oriental medical concepts or to modern techniques of diagnostic evaluation.

State laws may also specifically authorize acupuncture licensees to employ a wide panoply of AOM therapies such as moxibustion, cupping, Oriental or therapeutic massage, therapeutic exercise, electro-acupuncture, acupressure, dietary recommendations, herbal therapy, injection and laser therapy, ion cord devices, magnets, *qi gong,* and massage. The treatment of animals, the ordering of western diagnostic tests and the use of homeopathy are, with limitations, within the scope of practice in some states.

Referral practices

AOM practitioners receive training in biomedicine, including training on appropriate referrals to medical practitioners. Some states specify the process for referral, but in most cases medical referral is determined by the AOM practitioner. Conversely, patients seeking AOM practitioners are mostly self-referred or referred by word of mouth.

Patients may be referred to professional acupuncturists for a variety of complaints. In 2003, a World Health Organization study, *Acupuncture: Review and Analysis of Reports on Controlled Clinical Trials,* cited over 43 conditions that are treatable with acupuncture. While the list below is not comprehensive, it represents the typical array of complaints for which referral to an acupuncturist is quite appropriate:

- pain management, including musculoskeletal pain, rheumatologic or oncological pain, headache/migraine, and pain palliation at end of life

- neurogenic pain, including neuropathy associated with diabetes mellitus, chemotherapy, and the neuropathy associated with antiretroviral therapy
- women's health, including dysmenorrhea, premenstrual syndrome, amenorrhea, infertility, and pain associated with endometriosis
- nausea and vomiting associated with pregnancy, chemotherapy-induced nausea, and postoperative nausea
- symptoms associated with virally mediated disorders, including HIV, hepatitis B and C viruses, and the pain associated with herpes zoster and herpes simplex I and II
- neurological disorders, including Bell's palsy, cerebral vascular accident, and multiple sclerosis
- routine infections such as colds and flu
- dermatological complaints, such as dermatitis and Psoriasis
- gastrointestinal complaints, including gastralgia, inflammatory bowel disease, irritable bowel syndrome, and gastroesophageal reflux disease
- respiratory complaints, including mild to moderate asthma, shortness of breath, and cough
- urogenital complaints, including dysurea, cystitis, and impotence
- mild to moderate anxiety and depression
- chemical dependency, including abuse of opiates and sympathomimetics
- fatigue secondary to chronic illness, medical treatment, or surgery
- health issues related to aging and imbalance: kidney disease, heart disease, diabetes
- cancer-related issues such as pain and the side-effects of chemotherapy

Third-party payers

Increasingly, acupuncture is covered in many insurance plans, is reimbursed by other third-party payers for certain conditions, and is a part of many employer programs. For the most part, however, payments to individual providers are made by the patient. Indeed, AOM has grown within the US through word of mouth and referrals based on successful treatments and is increasingly recognized as a safe, effective medicine in a wide variety of settings including general health care, integrative medical practices, hospitals, spas and medispa practices, sports and fitness gyms, nonprofit community clinics, and addiction and recovery programs.[5,6,7] There remains, however, a significant disparity in traditional insurance coverage compared to other healthcare professions that have been established in the mainstream world of insurance reimbursement for several decades.

Legislation requiring the coverage of acupuncture services under federal Medicare and the Federal Employee Benefits Program (FEBP) has been introduced in Congress for a number of years. The enactment of this legislation has long enjoyed significant support from within and outside of the profession. Of the 12 insurance plans in the FEBP (see Resources for list of plans), eleven (all but Aetna), offered acupuncture benefits in 2012 covering a significant portion of the federal workforce. Additionally, there is increasing utilization of acupuncture by programs funded by the Department of Defense for active duty personnel and by the Department of Veterans Affairs, and some states have included acupuncture as an Essential Health Benefit.

Integration Activities

Interest in training that prepares AOM students to work in integrative healthcare settings is high among Council of Colleges of Acupuncture and Oriental Medicine (CCAOM)'s 51 AOM member institutions. A substantial number of these colleges offer internships in off-site clinics where AOM services are provided in local communities. These clinics include a variety of settings, including hospitals, multi-specialty centers, research-based centers, public health clinics, long- and short-term rehabilitation centers, family practice clinics, nursing homes, outpatient geriatric or assisted living centers for seniors, drug treatment centers, HIV/AIDS treatment facilities, sports medicine clinics, pediatric, cancer, and other specialty care centers, and clinics addressing specific community group needs, such as women's health and inner city/low income/multi-racial groups.

In addition, CCAOM, NCCAOM, and the Accreditation Commission for Acupuncture and Oriental Medicine (ACAOM) actively participate in the work of the Academic Consortium for Complementary and Alternative Health Care (ACCAHC) whose work, in collaboration with leaders from conventional medical educational institutions, has focused on developing strategies for integrating conventional and CAM education.

Education

Schools and programs

The first acupuncture school in the US was established in 1975. In 1982 the Council of Colleges of Acupuncture and Oriental Medicine (CCAOM) was founded as a nonprofit, voluntary membership association for AOM colleges and programs. Currently the membership of the CCAOM consists of 51 member schools, all of which have achieved accreditation, or are candidates for accreditation with ACAOM, the national accrediting agency for AOM schools. This membership is spread over 20 states, with major concentrations of colleges in California, New York, and Florida. There is significant diversity in the programs among CCAOM's member schools with representation of the Traditional Chinese Medicine, Japanese,

Five Element, Korean, and Vietnamese traditions. The growth of CCAOM since the early years of its founding, when there were fewer than 12 member schools, has mirrored the general growth of the profession. The mission of CCAOM is to advance AOM by promoting educational excellence in the field.

One of the goals supporting the mission of CCAOM is to take a leadership role in acupuncture safety through publication, education, and delivery of a national needle safety course. Pursuant to this goal, the Council's Clean Needle Technique (CNT) course is taught to AOM students to ensure that there is a national standard of needle safety protocol in the interest of public safety. Verification of successful completion of the Council's CNT course, as evidenced by issuance of a certificate by the Council, is required for a person to receive Diplomate status from NCCAOM.

The Executive Committee of the CCAOM consists of eight officers who are full-time presidents, CEOs, program directors, or other administrators at CCAOM's member schools. The full CCAOM meets twice each year in conferences that provide a forum for member institutions to dialogue about issues of importance to the colleges. In addition, CCAOM's meetings provide an opportunity for its many committees to meet face-to-face and for the membership to benefit from panel discussions, strategic planning, and specialized workshops.

Curriculum content

The entry-level standard of training and education for the profession is the master's degree, which is typically a 3-4 year program that offers either the first-professional master's degree or the first-professional master's level certificate or diploma. Under ACAOM's national accreditation standards, AOM institutions that offer a master's degree in acupuncture must be at least 3 academic years in length and provide a minimum of 1905 hours of a professional acupuncture curriculum over a three-year period (typically taken over four calendar years). The curriculum for an acupuncture program must consist of at least 705 didactic hours in Oriental medical theory, diagnosis, and treatment techniques in acupuncture and related studies; 660 hours in clinical training; 450 hours in biomedical clinical sciences; and 90 hours in counseling, communication, ethics and practice management.

A professional Oriental medicine curriculum, which is a minimum of four academic years and includes training in acupuncture and the additional study of Chinese herbal medicine, must consist of at least 2625 hours leading to a master's degree or master's level certificate or diploma in Oriental medicine. The curriculum for an Oriental medicine program must consist of at least 705 hours in Oriental medical theory, diagnosis, and treatment techniques in acupuncture and related studies; 450 hours in didactic Oriental herbal studies; 870 hours in integrated acupuncture and herbal clinical training; 510 hours in biomedical clinical sciences; and 90 hours in counseling, communication, ethics, and practice management.

For some years, there has been an ongoing dialogue within the profession concerning the amount of academic training a person needs to practice AOM at the entry level and what the most appropriate professional title should be for providers. This dialogue has been formally focused within a doctoral task force established by ACAOM, whose work is discussed in more detail in this chapter under the heading of Accreditation. The average number of academic hours of study and training associated with the entry-level Master's degree among AOM colleges nationally is between 2,700 and 2,900 hours. A significant number of Oriental medicine programs have a curriculum of 3,000 hours or more. The trend over the years has been for an increase in the number of academic hours.

As of the May of 2013, ACAOM had approved nine colleges to offer postgraduate clinical doctoral programs on a pilot basis (six have achieved accreditation and three had candidacy status for this degree). The degree for this program, titled Doctorate in Acupuncture & Oriental Medicine (DAOM), provides advanced training in either acupuncture or in Oriental medicine. DAOM programs must provide advanced didactic and clinical training in one or more clinical specialty areas in AOM. Completion of a master's degree or master's level program in acupuncture or in Oriental medicine is a prerequisite for admission into a DAOM program. Such programs must comprise a minimum of 1200 hours of advanced AOM training in one or more clinical specialty areas at the doctoral level. As of December 31, 2012, a majority of the faculty must possess a doctoral degree, the terminal degree, or its international equivalent in the subject areas in which the faculty teach. Clinical supervisors should have a minimum of five years of documented professional experience as licensed Oriental medicine practitioners.

There has been an ongoing discussion within stakeholder groups regarding a future first-professional doctorate for the acupuncture profession in the US. After a lengthy development process, ACAOM published competency based standards in March 2013 and began accepting substantive change applications for review in June 2013.

Faculty and other training information

The accreditation standards for master's level programs as prescribed by ACAOM have general requirements for AOM faculty. Thus, AOM faculty must be academically qualified and numerically sufficient to perform their responsibilities. Additionally, the general education, professional education, teaching experience, and practical professional experience of faculty must be appropriate for the subject area taught. Faculty members must also provide continuing evidence of keeping abreast of developments in the fields and subjects in which they teach. An increasing number of faculty members at AOM institutions now have doctoral degrees. This transition has been facilitated by institutions offering the post-graduate DAOM program.

Accreditation

ACAOM was established in 1982 and is the only national accrediting body recognized by the US Department of Education as a reliable authority for quality education and training in AOM. Its current scope of recognition with the USDE is the "...accreditation and pre-accreditation ("Candidacy") throughout the US of first-professional master's degree and professional master's level certificate and diploma programs in AOM, and professional post-graduate doctoral programs in acupuncture and in Oriental medicine (DAOM), as well as freestanding institutions and colleges of AOM that offer such programs...." As an accrediting agency, ACAOM's primary purposes are to establish comprehensive educational and institutional requirements for acupuncture and Oriental medicine programs; to accredit programs and institutions that meet these requirements; and to foster excellence in AOM through the implementation of accreditation standards for AOM educational institutions and programs.

ACAOM was first recognized by the US Department of Education (USDE) in 1988 for the accreditation of master's degree and master's-level acupuncture programs. In 1992, ACAOM was granted an expansion of scope by the USDE to include the accreditation of programs in Oriental medicine, which are programs that include the study of Chinese herbal medicine. In May 2006, the USDE renewed ACAOM's recognition for USDE's maximum five-year period.

Effective July 8, 2011, ACAOM received continued recognition by USDE for one year. ACCAOM submitted a compliance report to USDE on August 8, 2012 and is scheduled for review at the June 2013 National Advisory Committee on Institutional Quality and Integrity (NACIQI) meeting.

In conjunction with the July 8, 2011 continued recognition from USDE, the scope of ACAOM's authority was expanded to include candidacy and accreditation of DAOM programs. Schools that have institutional accreditation from ACAOM may now use that accreditation to establish eligibility for Title IV funding for DAOM programs.

The establishment of ACAOM and its recognition by USDE have made it possible for AOM students to obtain federal student loans for their education. When the USDE renewed ACAOM's recognition in 2006, the federal agency also granted ACAOM's request for an expansion of scope to include candidacy reviews, thus making it possible for nonprofit, freestanding AOM institutions that have achieved candidacy status to establish eligibility for their students to participate in federal Title IV student aid programs; however, USDE's Title IV regulations require for-profit or proprietary AOM institutions to achieve accreditation to establish eligibility for their students to participate in Title IV student aid programs.

ACAOM has over 60 schools and colleges with accreditation or candidacy status. Additional AOM schools that have not achieved candidate or accreditation status with ACAOM do exist in the US.

The recent national trend toward the provision of healthcare services in integrated CAM and conventional settings is reflected in the work of ACAOM's Doctoral Task Force (2003-2012), which was composed of representatives from a broad segment of the AOM profession. The task force drafted outcomes and competency-based accreditation standards for the first-professional entry-level doctorate that take into account the increasing acceptance of AOM practice in conventional and complex medical environments. The task force completed its review of public comments in August of 2012 and forwarded final recommendations to ACCAOM.

Regulation and Certification

Regulatory status

Currently, the right to practice acupuncture by comprehensively trained AOM practitioners exists in 44 states and in the District of Columbia as described in each state's practice act. The right to practice may be designated by licensure, certification, or registration under the applicable state law. Licensure is the most common form of authorization to practice. In states without regulation, practitioners typically practice subject to potential oversight from a medical board. Other states may limit practice specifically to designated medical providers or AOM practitioners who are medically supervised. In the few remaining unregulated states, activities to obtain full licensure status for acupuncturists are under way.

In most states with regulation, professional acupuncturists have independent status, although there remain a few states where practitioners must have supervision, prior referral, or initial diagnosis by a conventional medical doctor. The recent statutory trend, however, is in favor of more professional independence by AOM providers. The most common designation for comprehensively-trained practitioners is Licensed Acupuncturist (LAc), although in a few states they may be designated by statute as Acupuncturist Physicians (AP-Florida) or Doctors of Oriental Medicine (DOM-New Mexico and Nevada), but these doctoral designations are licensure titles conferred by the state and do not reflect earned academic degrees at the doctoral level. The State of Washington now specifies Eastern Asian Medicine Practitioner (EAMP) although the revised statute allows the continued use of Licensed Acupuncturist (LAc).

The state statutes regulating acupuncture are not uniform. In some states there are very detailed statutes and regulations, but in others there may be only a few paragraphs concerning the practice of acupuncture. The administrative structure for the regulation of acupuncture in the states also varies considerably. The most common structure is for the profession to be regulated by an independent board composed of professional acupuncturists or by a state medical board with the assistance of an advisory acupuncture board or committee. Other administrative arrangements include regulation by a joint board comprised of diverse healthcare professionals (both conventional and CAM), by the board of another CAM profession, or by a larger administrative division within a state department—all typically with the assistance of an acupuncture advisory body. This diversity reflects political and budgetary realities as each state tailors its law to meet local needs.

All states regulating the practice of acupuncture require or accept passage of NCCAOM certification examinations or, in the case of 23 states, require full NCCAOM certification as either a Diplomate of Acupuncture (Dipl. Ac. [NCCAOM]) or Diplomate of Oriental Medicine (Dipl. O.M. [NCCAOM]). Currently, the only state not using the NCCAOM certification exams for licensure is California. California provides its own licensing exam. Figure 1.1 illustrates those states that use NCCAOM examinations or require full certification by NCCAOM.

In general, national AOM organizations have been supportive of the adoption of state acupuncture practice acts that incorporate reference to national standards of education, training, and certification in the field. Adherence to such standards, as administered by ACAOM in the field of accreditation and by NCCAOM for certification, assures a high level of practitioner competence and a degree of uniformity that facilitates portability among the various states in the recognition of practitioner credentials.

Figure 1.1

States that Recognize NCCAOM Certification or Examination for Licensure
States that Recognize NCCAOM Certification or Examination for Licensure

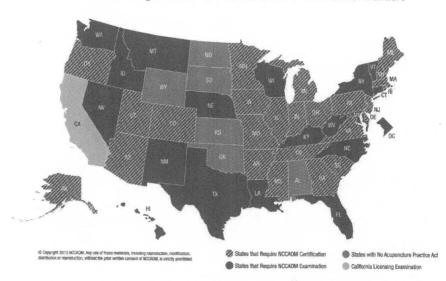

© Copyright 2013 NCCAOM. Any use of these materials, including reproduction, modification, distribution or reproduction, without the prior written consent of NCCAOM, is strictly prohibited.

States that Require NCCAOM Certification   States with No Acupuncture Practice Act
States that Require NCCAOM Examination   California Licensing Examination

Currently, acupuncture is not covered under the federal Loan Repayment Program. In recent years, the profession has expressed interest in and sought to obtain such coverage. The list of eligible providers under the Program, however, is essentially limited to those within the conventional healthcare field.

Examinations and certifications

Graduates of ACAOM accredited or candidate institutions are qualified to take the national certification exams offered by NCCAOM. Established in 1982, the mission of NCCAOM is to establish, assess, and promote recognized standards of competence and safety in acupuncture and Oriental medicine for the protection and benefit of the public. NCCAOM is a member of the Institute for Credentialing Excellence (ICE) and its certification programs in Acupuncture, Chinese Herbology, and Oriental Medicine are accredited by the National Commission for Certifying Agencies (NCCA), a separate independent commission of ICE. Accreditation by NCCA ICE represents the highest voluntary certification standards in the US. Passage of two or more of NCCAOM's national examinations are a requirement of licensure recognized in 43 states and the District of Columbia. Only California has its own exam; however, movement is growing in California to have the NCCAOM certification exams accepted there. Candidates who pass NCCAOM's required certification exams in Acupuncture, Chinese Herbology, Oriental Medicine or Asian Bodywork Therapy are awarded the designation NCCAOM *Diplomate* appropriate to the certification achieved: Dipl. Ac. (NCCAOM), Dipl. O.M. (NCCAOM), Dipl. C.H. (NCCAOM), and Dipl. ABT (NCCAOM).

The first NCCAOM comprehensive written examination in acupuncture was administered in 1985 and was developed over a three-year period with the assistance of leading acupuncturists throughout the US. In 1989, NCCAOM added a practical examination of point location skills as a component of its acupuncture examination. A clean needle technique exam was added to the certification requirements for the acupuncture written exam in 1991 and merged into the acupuncture exam in 1998. The NCCAOM administered the first national examination in Chinese Herbology in 1995 and in 2000 offered its first written examination in Asian Bodywork Therapy. In 2003, NCCAOM began to offer an umbrella certification in Oriental medicine to applicants who demonstrated competence in both acupuncture and Chinese herbology, as well as entry-level competency in biomedicine.

Since its inception, the NCCAOM has issued more than 25,000 certificates in acupuncture, Oriental Medicine, Chinese Herbology, and Asian Bodywork Therapy and reports the existence of more than 16,000 active Diplomates worldwide in current practice. Figure 1.2 shows the number of actively certified (i.e. newly certified or recertified) NCCAOM Diplomates per state. The NCCAOM requires all active Diplomates to recertify every four years by submitting documentation of 60 professional development activity points.

Figure 1.2

Number of Active NCCAOM Diplomates Per State

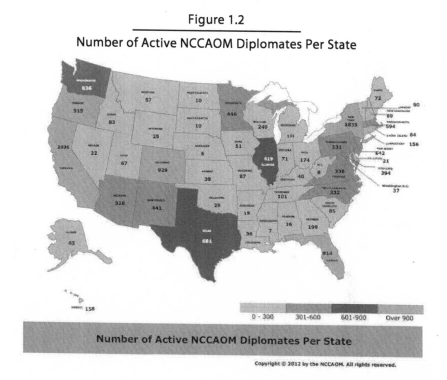

Number of Active NCCAOM Diplomates Per State

Research

Research in the US on the efficacy of AOM increased after 1996 when the FDA reclassified acupuncture needles from experimental devices (Class III) to devices for which performance standards exist (Class II). This provided reasonable assurance that acupuncture needles would be safe, thus making research easier to conduct. Subsequent investigations over the last decade have found that acupuncture needles had few side effects, making them somewhat safer than certain conventional western medical treatments for many diseases. Research in AOM has increased dramatically over the past decade within the US, as well as in Asia, Europe, and South America.

Among the more prominent scientific endorsements of the efficacy of acupuncture is that of the National Institutes of Health, which concluded in a 1997 Consensus Statement that:

> *...promising results have emerged, for example, showing efficacy of acupuncture in adult postoperative and chemotherapy nausea and vomiting and in postoperative dental pain. There are other situations such as addiction, stroke rehabilitation, headache, menstrual cramps, tennis elbow, fibromyalgia, myofascial pain, osteoarthritis, low back pain, carpal tunnel syndrome, and asthma, in which acupuncture may be useful as an adjunct treatment or an acceptable alternative or be included in a comprehensive management program. Further research is likely to uncover additional areas where acupuncture interventions will be useful.*[8]

Within the AOM profession, the most prominent research organization in the US is the Society for Acupuncture Research (SAR), which was formally established in 1993 and whose mission is to promote, advance, and disseminate scientific inquiry into Oriental medicine systems, which include acupuncture, herbal therapy and other modalities. SAR values quantitative and qualitative research addressing clinical efficacy, physiological mechanisms, patterns of use, and theoretical foundations. The organization sponsors biennial symposia on research methodologies and welcomes individual affiliates including researchers, educators, students, acupuncturists, healthcare practitioners, and members of the public, as well as institutional affiliates including schools, vendors, and other organizations to participate in these events.

In November 2007, SAR held an international symposium aimed at presenting and discussing the progress made in acupuncture research during the decade following the NIH Consensus Development Conference. The symposium presentations, as well as their summaries are published in the *Journal of Alternative and Complementary Medicine*.[9,10,11]These reports unequivocally show that the field of acupuncture research has significantly expanded and matured since 1997. Phase II/III sham-controlled trials have been successfully completed and a broad range of basic research studies have identified numerous biochemical and physiological correlates of acupuncture. However, SAR has also identified intriguing paradoxes emerging from the symposium and summaries that touch on some disconnects, including those between clinical and basic acupuncture research. A discussion of these issues was published in 2011.[12]

The international conference of SAR, held in 2010, was thematically organized around the role of translational research within the field of acupuncture. More than 300 acupuncture researchers, practitioners, students, funding agency personnel and health policy analysts from twenty countries attended this meeting. SAR intends to hold biennial conferences starting with the 2013 international conference and will serve as a venue for timely reporting of the latest acupuncture research highlights destined for high impact research journals. These conferences are also an excellent forum for strengthening research curricula in CCAOM's member schools. In 2011, CCAOM established a pilot research grant program to encourage collaborative research proposals and improve the research capacity of its member colleges.

Challenges and Opportunities

Key challenges 2013 – 2018

- need for greater consensus concerning what primary care competencies are appropriate for LAc's
- need for greater public awareness of the skill limitations of other healthcare providers who practice acupuncture without having completed a full three- or four-year AOM curriculum at an accredited AOM institution or program, but instead take a weekend or continuing education course in acupuncture, or only complete 200 hours or less of academic training in acupuncture
- need for greater reciprocity among states in recognizing the credentials of AOM providers
- need for an appropriate response to increasing professional and public expectations for integrated health care
- regulatory uncertainty concerning FDA restrictions on the use of Chinese herbs
- need for reimbursement for AOM treatments by HMOs, Medicare, and third-party payers, and for identification of practitioner qualifications/standards for obtaining such reimbursement
- need for greater emphasis on AOM research, training of AOM researchers, and increased funding for AOM research
- need to strengthen and coordinate national representation and AOM profession's voice for federal policy issues
- need for greater public awareness and acceptance of AOM with associated increase in viable professional practice opportunities for a greater number of AOM practitioners many of whom graduate with substantial student loan debt
- need for greater awareness of the qualifications of licensed acupuncturists in the provider credentialing process at hospitals and major medical centers

- need for more comprehensive and coordinated collection of data concerning the profession

Key opportunities 2013 – 2018

- work of ACAOM to implement the first-professional doctorate (FPD) and review/revise the existing Master Standards consistent with the new competency based standards
- increase in number of AOM schools offering postgraduate clinical education through the Doctorate of Acupuncture and Oriental Medicine programs
- work of ACCAHC focusing on interdisciplinary CAM and conventional healthcare education
- growing commitment within the profession for greater collaboration within the field, including joint projects among the national and state organizations
- increased marketing activities to develop public awareness of the benefits, safety, and cost-effectiveness of AOM
- success and visibility of emergency acupuncture services provided in aftermath of Hurricane Katrina (and other national disasters) which are a precedent for similar opportunities in future emergency situations and for greater coordination within the profession for local emergency response
- initial efforts underway to obtain greater recognition of acupuncture as a separate profession under the federal Standard Occupational Classification (SOC) through the BLS
- expansion of education and accreditation to include advanced training in AOM areas of bodywork such as *tui na* and *shiatsu*, nutrition, meditation, and exercise.

Resources

Organizations and websites

- Accreditation Commission for Acupuncture and Oriental Medicine (ACAOM) www.acaom.org
- Acupuncturists Without Borders (AWB) www.acuwithoutborders.org
- American Association of Acupuncture and Oriental Medicine (AAAOM) www.aaaomonline.org
- American Organization for Bodywork Therapies of Asia (AOBTA) www.aobta.org
- Council of Colleges of Acupuncture and Oriental Medicine (CCAOM) www.ccaom.org
- National Acupuncture Detoxification Association (NADA) www.acudetox.com
- National Certification Commission for Acupuncture and Oriental Medicine (NCCAOM) www.nccaom.org
- Society for Acupuncture Research (SAR) www.acupunctureresearch.org

Federal Employee Benefits Program (FEBP) Plans

- Aetna Health Fund (nearly nationwide) http://www.opm.gov/insure/health/planinfo/2012/brochures/73-828.pdf
- APWU Health Plan http://www.opm.gov/insure/health/planinfo/2012/brochures/71-004.pdf
- BCBS Service Benefit Plan http://www.opm.gov/insure/health/planinfo/2012/brochures/71-005.pdf
- Compass Rose Health Plan (Restricted enrollment based on agency)

http://www.opm.gov/insure/health/planinfo/2012/brochures/72-007.pdf

- Foreign Service Benefit Plan (Restricted enrollment based on agency)
  http://www.opm.gov/insure/health/planinfo/2012/brochures/72-001.pdf
- GEHA Benefit Plan (High and Standard)
  http://www.opm.gov/insure/health/planinfo/2012/brochures/71-006.pdf
- GEHA High Deductible Health Plan
  http://www.opm.gov/insure/health/planinfo/2012/brochures/71-014.pdf
- Mail Handlers High Deductible (Consumer Option) Health Plan
  http://www.opm.gov/insure/health/planinfo/2012/brochures/71-016.pdf
- Mail Handlers Benefit Plan (Standard and Value)
  http://www.opm.gov/insure/health/planinfo/2012/brochures/71-007.pdf
- NALC Health Benefit Plan
  http://www.opm.gov/insure/health/planinfo/2012/brochures/71-009.pdf
- Rural Carrier Benefit Plan (enrollment restricted to active/retired letter carriers)
  http://www.opm.gov/insure/health/planinfo/2012/brochures/72-005.pdf
- SAMBA
  http://www.opm.gov/insure/health/planinfo/2012/brochures/71-015.pdf

Bibliography:

Bensky D, Barolet R, Ellis A, Scheid V. *Chinese Herbal Medicine Formulas and Strategies*. 2nd ed. Seattle, WA: Eastland Press; 2009.

Bensky D, Clavey S, Stroger E. *Chinese Herbal Medicine Materia Medica*, 3rd ed. Seattle, WA: Eastland Press; 2004.

Chen J, Chen T. *Chinese Herbal Formulas and Applications: Pharmacological Effects and Clinical Research*. City of Industry, CA: Art of Medicine Press; 2009.

Chen J, Chen T. *Chinese Herbal Medicine and Pharmacology*. City of Industry, CA: Art of Medicine Press; 2003.

Cheng X, ed. *Chinese Acupuncture and Moxibustion*. Beijing, China: Foreign Languages Press; 1987.

Deadman P, Al-Khafaji A. *A Manual of Acupuncture*. Sussex, England: Journal of Chinese Medicine Publications; 1998.

Huang Ti Nei Ching Su Wen. Veith I, trans. Berkeley, CA: University of California Press; 1949.

Maciocia G. *The Foundations of Chinese Medicine*. Philadelphia, PA: Churchill Livingstone; 1989.

Unschuld P, trans. *Nan Ching*. Berkeley, CA: University of California Press; 1986.

Unschuld P. *Huang Di Nei Jing Su Wen: Nature, Knowledge, Imagery in an Ancient Chinese Medical Text*. Berkeley, CA: University of California Press; 2003.

Unschuld P. *Medicine in China: A History of Ideas*. Berkeley, CA: University of California Press; 1988.

Wiseman N, Ye F. *A Practical Dictionary of Chinese Medicine*. Brookline, MA: Paradigm Publications; 1998.

Zmiewski P, Wiseman N, Ellis A, eds. *Fundamentals of Chinese Medicine*. Brookline, MA: Paradigm Publications; 1985.

Citations

1. Reston J. Now let me tell you about my appendectomy in Peking. *New York Times*. July 26, 1971
2. Ward-Cook K and Hahn T, eds. 2008 Job Task Analysis: A Report to the Acupuncture and Oriental Medicine Profession, 2009, NCCAOM. www.nccaom.org.

3. ACAOM Reports Enrollment and Graduation Numbers. Feb. 29,2012. http://www.acaom.org/documents/acaom_reports_enrollment_and_graduation_numbers.pdf. , accessed 2/20/2013.

4. Jabbour M, Sapko MT, Miller DW, et al. Economic evaluation in acupuncture: past and future, the *American Acupuncturist.* 2009; Fall(49):11.

5. Xu S, Want L, Cooper E, et al. Adverse events of acupuncture: a systematic review of case reports. *Evidence-Based Complementary and Alternative Medicine.* 2013, Vol. 2013: 1-15.

6. Witt CM, Jena S, Brinkhaus B, et al. Acupuncture in patients with osteoarthritis of the knee or hip: a randomized, controlled trial with an additional nonrandomized arm. *Arthritis Rheum.* 2006, Nov; 54(11):3485-93.

7. Cherkin DC, Sherman KJ, Avins AL, et al. A randomized trial comparing acupuncture, simulated acupuncture, and usual care for chronic low back pain. *Arch Intern Med*, 2009;169(9):858-866.

8. Acupuncture. NIH Consensus Statement 1997 Nov 3-5; 15(5):1-34.

9. MacPherson H, Nahin R, Paterson C, Cassidy CM, Lewith GT, Hammerschlag R. Developments in acupuncture research: big-picture perspectives from the leading edge. J Altern Complement Med. 2008;14(7):883-7.

10. Park J, Linde K, Manheimer E, Molsberger A, Sherman K, Smith C, Sung J, Vickers A, Schnyer R. The status and future of acupuncture clinical research. J Altern Complement Med. 2008;14(7):871-81.

11. Napadow V, Ahn A, Longhurst J, Lao L, Stener-Victorin E, Harris R, Langevin HM. The status and future of acupuncture mechanism research. J Altern Complement Med. 2008 Sep;14(7):861-9. PMID: 18803495

12. Langevin H, Wayne P, MacPherson H, et al. Paradoxes in acupuncture research: strategies for moving forward, *Evidenced Based Complementary Alternative Medicine*, 2011, doi:10.1155/2011/180805.

Chiropractic

First Edition (2009) Authors: Reed Phillips, DC, PhD;
Michael Wiles, DC, MEd, MS; David O'Bryon, JD, CAE

Second Edition (2013) Editors: Michael Wiles, DC, MEd,
MS; David O'Bryon, JD, CAE; Joseph Brimhall, DC

Partner Organization: Association of Chiropractic Colleges

About the Authors/Editors: Phillips is former President of the
Southern California University of Health Sciences, past Vice
President of the Foundation for Chiropractic Education and
Research, and a past President of both the Association of
Chiropractic Colleges and the Council on Chiropractic
Education. Wiles is Provost and Vice President for Academic
Affairs of Northwestern Health Sciences University and co-chair
of the ACCAHC Education Working Group. O'Bryon is
Executive Director of the Association of Chiropractic Colleges,
Vice Chair of ACCAHC, past President of the Federation of
Associations of Schools of the Allied Health Professions and
serves on the Secretariat Board of the National Association of
Independent Colleges and Universities. Brimhall is the President
of the University of Western States, a member of the ACCAHC
Executive Committee, immediate past President of the Councils
on Chiropractic Education International, and a former Board
President and Commission Chairman of the Council on
Chiropractic Education.

Philosophy, Mission, Goals

Chiropractic is a healthcare discipline that emphasizes the
inherent power of the body to heal itself. A Doctor of Chiropractic practicing primary health care is competent and qualified

to provide independent, quality, patient-focused care to individuals of all ages and genders by: 1) providing direct access, portal-of-entry care that does not require a referral from another source; 2) establishing a partnership relationship with continuity of care for each individual patient; 3) evaluating a patient and independently establishing a diagnosis or diagnoses; and, 4) managing the patient's health care and integrating healthcare services including treatment, recommendations for self-care, referral, and/or co-management.[1]

The above definition is taken from January 2012 Accreditation Standards from the Council on Chiropractic Education (CCE), the chiropractic accrediting agency recognized by the United States Department of Education. This definition is used by Doctor of Chiropractic degree programs in establishing educational curricula and clinical competencies that prepare graduates for professional chiropractic practice. Preferences and practices vary in how the profession describes itself.

A Doctor of Chiropractic may, depending on jurisdictional definitions and individual preference, utilize any of a range of titles: chiropractor, doctor of chiropractic, doctor of chiropractic medicine, or chiropractic physician. Some will only use "chiropractic" to describe the practice while others prefer "chiropractic medicine." This range of usages is also found in statutory and regulatory language. In some jurisdictions, laws establish the terms and titles that are used. The Department of Labor description includes the terms "chiropractic physician" and "chiropractic medicine."

Inside the Association of Chiropractic Colleges (ACC), individual member institutions may also choose diverse designations. For this reason we will intersperse mixed use of these terms in this chapter. We urge practitioners from other disciplines to engage the members of the chiropractic profession on their legally-established titles and preferences. This may provide a useful way to improve understanding and collaboration.

Early chiropractic concepts of health and disease were compatible with vitalistic philosophies of the late 19th century. Living cells were conceived as having an inborn intelligent component that was responsible for the maintenance of life. This phenomenon is analogous to homeostasis. Some homeostatic control mechanisms function as a result of coordination of sensory input and motor output through the central nervous system. The intimacy of body structure (particularly the spine) and the nervous system, and hence body function, led the early chiropractors to develop a philosophy of care that, in essence, posited that a normally functioning nervous system, in the presence of structural normality, should lead to normal health. This vitalistic approach toward health and disease was similar to that proposed by the early osteopathic profession.

History of the profession

Chiropractic was founded by Daniel David Palmer in Davenport, Iowa in 1895. Palmer was originally a schoolteacher who developed an interest in magnetism and what was, at that time, popularly called magnetic healing. Having observed that a patient's deafness had apparently occurred following a traumatic experience resulting in a protuberance on his spine, Palmer reasoned that reduction of this protuberance might affect the patient's hearing. Evidently, following a manipulation of the spine, the patient's hearing improved. This led to the form-ulation of a theory relating spinal alignment to states of health and disease which was similar to that espoused by Andrew Taylor Still, the founder of osteopathy, in 1874 in neighboring Missouri. In 1975, the National Institutes of Health sponsored an interdisciplinary conference called "The Research Status of Spinal Manipulative Therapy" in Bethesda, Maryland. This conference brought together leaders in the field of spinal manipulation from the chiropractic, medical, and osteopathic professions from around the world. It marked a significant point in the evolution of chiropractic education and practice, and important progress has been made since that time toward the integration of chiropractic education, science, research and practice into mainstream systems of health care and professional education. While this process is far from complete, the chiropractic profession of today can be proud of the great advances in education and research of the past few decades.

Chiropractic spread first to Canada, then to the United Kingdom, and finally throughout the world from these simple roots in Davenport, Iowa. Educational institutions offer chiropractic degrees in Europe, Africa, Australia, Asia, and North and South America, with new programs currently under development. The Councils on Chiropractic Education International (CCEI) has published the International Chiropractic Accreditation Standards to help establish the highest possible quality in chiropractic education programs worldwide. The World Health Organization has also published guidelines on basic training and safety in chiropractic education and practice.[2]

Characteristics and Data

The chiropractic medical profession is growing rapidly throughout the world. There are approximately 72,000 chiropractors in the US and about another 16,000 in the rest of the world. Doctors of chiropractic are regulated as members of a licensed healthcare profession in all jurisdictions of the United States and Canada, in Mexico, and in about 40 other countries. In roughly 60 additional countries, chiropractors practice in an unregulated fashion, and many of these countries are currently in the process of officially recognizing and regulating the practice of chiropractic health care. In countries where chiropractic medicine is regulated, many of the chiropractic degree programs and colleges are located within university settings. The CCEI and the International Board of Chiropractic

Examiners are working with the as yet unregulated countries to establish and maintain consistent and equivalent standards of education and examination.

Income data vary for chiropractors, depending on the country and source of the data. Most sources suggest that full-time chiropractic physicians earn an income comparable with other health professionals. In two recent annual salary and expense surveys, chiropractors reported an average total compensation of $107,000 in 2012 and $123,375 in 2011.[3,4]

Clinical Care

Approach to patient care
Today, the traditional vitalistic philosophy has evolved to a patient care approach that recognizes and honors the body's own innate mechanisms for adaptation and homeostasis. Treatment is designed to optimize and support the body's natural self-healing and intrinsic regulatory systems.

In practical terms, this can be translated into an approach to care that seeks to relieve symptoms, restore joint motion, enhance posture and balance, provide necessary support, strengthen muscles, improve flexibility and facilitate coordination to optimize body function. A patient may seek help for lower back pain, for example, and this approach to care will address not only the local factors associated with the back pain, but also the structural adaptations that may have led to it in the first place, such as weak muscles, obesity, detrimental postures, and other factors. Diet, activities of daily living, lifestyle behaviors and stress reduction may also be addressed.

Such an approach can be considered both patient-centered and holistic. Chiropractic physicians generally approach patient care in a manner that is similar to conventional medical doctors. Procedurally, the patient is interviewed, a detailed health history is obtained, an examination is performed including any necessary specialized tests, results are compiled and reviewed, and a working diagnosis is formulated. Then, a comprehensive management plan is constructed and recommended, and the patient is afforded an opportunity to provide informed consent before initiating any treatment. Clinical progress is monitored, and the patient is discharged from active care when the appropriate outcomes have been achieved.

Chiropractic physicians establish a standard medical history and are particularly concerned with the identification of factors or conditions that may require either referral to, or co-treatment with, other healthcare providers. Patients commonly seek chiropractic services for complaints related to the musculo-skeletal system; accordingly, musculoskeletal diagnoses are common in chiropractic practices.

Doctors of chiropractic focus not only on the structural component of a patient's complaint, but also evaluate the overall health of the patient. For example, a lower back complaint may be related locally to a sacro-iliac joint dysfunction, and generally to poor posture and obesity, all of which must be taken into consideration in the comprehensive treatment and management plan.

Clinical management plans are developed to provide symptom relief and achieve problem resolution, and to optimize whole-person health and function. Typically, treatment involves manual therapy, often including spinal manipulation, and other forms of treatment are commonly used as well. These may include exercise, stretching, rehabilitative measures, physical therapeutics (such as electrotherapy, hydrotherapy, or ultra-sound), diet and nutritional counseling, lifestyle advice, and recommendations for stress reduction. Manual therapy includes a continuum of treatment methods ranging from stretching and sustained pressure techniques to specific joint manipulation, often referred to as chiropractic "adjustments". Some techniques employ articulated treatment tables, and some use mechanical devices that move, stretch or position body structures. There is a very wide spectrum of techniques and methods under the umbrella of manual therapies, and the one most commonly associated with chiropractic treatment is the chiropractic manipulation or "adjustment" typically delivered by hand with a very specific and quick, gentle thrust.

The purpose of the chiropractic adjustment is to improve joint motion and function. The theoretical model is a dysfunctional articular lesion, sometimes traditionally or historically referred to as a subluxation. Joint dysfunction may be associated with a correctible segmental mechanical phenomenon characterized by asymmetry, restriction of motion, tissue texture abnormalities, and tenderness to palpation. Neurologically, spinal joint dysfunction, or subluxation, has been associated with segmental facilitation, and with trophic changes. This "manipulable lesion," as some have called it, is widely known to practitioners of manual medicine and spinal manipulation but is not generally known or appreciated by practitioners outside of those fields. Early chiropractic theories of a "bone out of place pinching a nerve" have plagued the profession for decades and have hindered interprofessional dialogue with regard to this lesion and phenomenon. A new generation of chiropractors and chiropractic educators are making strides toward standard-ization of terminology, the elucidation of the properties of segmental joint dysfunction, and the biomechanical effects of spinal manipulative therapy.

The majority of patients consult a chiropractor for complaints directly related to back pain and other musculoskeletal complaints. Patients also seek care from chiropractic physicians for headaches, digestive problems, hypertension, fatigue, and many other conditions that are typically seen in a primary healthcare setting.

Scope of practice

The scope of practice for doctors of chiropractic is established by jurisdictional laws and regulations, similar to all other licensed professions. Three features of legislation and practice are common in all jurisdictions that regulate chiropractic health care:

1. primary access, in which doctors of chiropractic accept patients directly without the requirement of referral from any other source
2. authority and obligation to establish a diagnosis prior to the initiation of treatment, which includes the authority to perform examinations and to order necessary diagnostic studies
3. authority to manage patient care by directly providing treatment, referring to another provider for additional care, and recommending lifestyle changes to facilitate health and wellness

A number of jurisdictions also allow chiropractic physicians to perform minor surgery, practice obstetrics, and prescribe medication. The scope of practice for chiropractic medicine varies widely among jurisdictions, and generally includes the diagnosis and treatment of human conditions. All manual treatment methods and the use of adjunctive approaches such as nutritional therapy, counseling, and physical therapeutics are also commonly included in chiropractic regulation.

Referral practices

Chiropractic students are taught the importance of proper and appropriate referral to other healthcare providers. Chiropractic medical education emphasizes the identification of conditions that may require specialized care and incorporates the protocols and processes for referring patients to practitioners that can provide such care. For example, since back pain can be due to a variety of diseases and medical conditions, some of which require urgent referral (such as cauda equina syndrome), doctors of chiropractic are trained to detect and manage patients with such conditions, including any necessary referral. Practicing chiropractic physicians seek to develop referral relationships with conventional medical physicians, osteopathic physicians, and naturopathic physicians, as well as specialists such as orthopedic surgeons, rheumatologists, neurologists, and gynecologists. Patient needs may also require referral to other practitioners, such as massage therapists, acupuncturists, physical therapists, occupational therapists, clinical psychologists, or other licensed and qualified healthcare practitioners.

Many doctors of chiropractic accept patients upon the referral of other healthcare providers. Specifically, patients with complaints related to the musculoskeletal system, such as back pain, neck pain, mechanical headaches, sports injuries, repetitive strain injuries, motor vehicle accident injuries, work-related musculoskeletal injuries, overuse syndromes, and other similar conditions, are frequently referred by other practitioners to chiropractic physicians for evaluation and treatment. Chiropractic physicians establish mutual referral networks with other physicians and healthcare providers. Chiropractors typically also develop a referral network with other chiropractic physicians, particularly those specializing in certain conditions. Chiropractic physicians are trained to establish and maintain records of patient care, and this information is routinely provided to other health practitioners that are involved in the care of a specific patient.

Third-party payers

Insurance coverage and third-party payment for chiropractic medical services vary widely from jurisdiction to jurisdiction. Chiropractic medicine is widely covered through private insurance plans in most countries, particularly in the US. Most government sponsored workers' compensation plans cover chiropractic services. There is an increasing tendency for chiropractic medicine to be included in wellness programs or other similar employer-sponsored health plans.

Integration Activities

All educational institutions offering chiropractic education have student clinics where the underserved or uninsured may receive chiropractic care at little or no cost. A growing number of community clinics include chiropractic medical services alongside the services of medical doctors and other healthcare services. Chiropractic physicians are increasingly involved in larger clinics and hospitals, and a growing number of hospitals grant privileges for chiropractors to treat patients on an outpatient basis and to use diagnostic facilities.

There are many other clinical settings where doctors of chiropractic are part of a multidisciplinary team. An outstanding example is the doctors of chiropractic who serve in the Walter Reed National Military Medical Center. Their presence at this facility is the outcome of ongoing negotiations with the US Department of Defense to employ Doctors of Chiropractic in military healthcare facilities as directed by Congress and the President of the United States.

Following recent legislation, chiropractic medicine is gradually becoming available throughout the military healthcare system. Experience to date has shown that chiropractic physicians in military healthcare facilities quickly become integral and valued members of the healthcare team.

With Presidential insistence, chiropractic medicine has also been introduced into the US Veterans Administration healthcare system. A three-year dialogue with a Federal Advisory Committee consisting of medical and osteopathic physicians, a physical therapist, a physician's assistant, and several doctors of chiropractic resulted in a list of 68 recommendations to the Secretary of the Veteran's Administration on the implementation process. Of the 68 recommendations, 67 were unanimously agreed upon. This program is slowly expanding throughout the entire VA system and current laws mandate the inclusion of chiropractic medical services at all VA facilities within the next few years. Several chiropractic educational institutions now have agreements with their local VA hospitals allowing clinical rotations of senior chiropractic students and interns, and facilitating interaction of conventional medical and chiropractic students, interns, and residents.

Chiropractic has become a highly accepted form of treatment in the world of sports. Most major professional teams in all sports employ a team of chiropractic physicians. Many college and university teams engage the services of a doctor of chiropractic as well. Doctors of Chiropractic provide care in many international events such as wrestling, track and field, swimming, and other sports, and are increasingly participating on multidisciplinary healthcare teams at the Olympics and in professional sports. For example, chiropractic physicians are on staff at the US Olympic Training Center in Colorado Springs, Colorado.

As the world of integrative medicine and complementary and alternative health care continues to grow, doctors of chiropractic medicine are at the forefront in many areas and are providing leadership to help establish standards and achieve cultural recognition of the benefits of complementary and alternative health care. Several chiropractic educational institutions have incorporated programs in acupuncture and Oriental medicine within their structure, at least one has added a program in naturopathic medicine, and many teach the principles and application of homeopathic formularies, clinical nutrition, life-style counseling, mind-body medicine, yoga, and massage therapy.

Over the last decade of increased integration activity, individual Doctors of Chiropractic and leaders of chiropractic programs have engaged various interprofessional and inter-institutional relationships with integrative medical programs that are among the 54 member programs of the Consortium of Academic Health Centers for Integrative Medicine. Among these are programs at Yale University, Harvard University, the University of Minnesota, and Oregon Health and Sciences University (OHSU). In the latter instance, the president of University of Western States Chiropractic College serves as a Director for the Oregon Collaborative for Integrative Medicine (OCIM), where his colleagues include academic leaders at OHSU and the presidents of the Oregon College of Oriental Medicine and the National College of Natural Medicine, all located in Portland, Oregon. The direction of growth and development is toward a more integrated practice among all healthcare disciplines.

Education

Schools and programs

Chiropractic colleges in the US grant the Doctor of Chiropractic degree after a course of study which consists of a minimum of 4200 classroom hours, typically delivered over a 4–5 academic year program. There are fifteen (15) educational institutions in the US and two (2) in Canada that offer Doctor of Chiropractic degree programs. These include individual chiropractic colleges, colleges within private universities, and a college within a public university system. Typically, chiropractic students are in their mid-twenties, with about 75% having completed undergraduate degrees before entering a chiropractic college.

At least 90 semester hours (three years) in undergraduate studies in the biological, physical and social sciences are required prior to admission to chiropractic medical programs in the United States and Canada. Outside North America, chiropractic degree programs are designed to be educationally similar to other direct-access, first-professional healthcare programs. After successfully completing the program of studies, the graduate earns a Doctor of Chiropractic (DC) degree in the US.

Outside of the US, graduates of chiropractic programs may receive other academic designations that are consistent with local jurisdictional requirements and customs. The most common of these credentials are the Bachelor of Science and Master of Science degrees in Chiropractic. In the United Kingdom, a Bachelor of Chiropractic (BChir) is awarded, analogous to the Bachelor of Medicine (MB or BMed) degree.

The Association of Chiropractic Colleges (ACC) represents all accredited colleges in the US and several others from around the world. Programs outside the US are most commonly affiliated with public universities. Accredited chiropractic degree programs exist in Canada (2), United Kingdom (2), Denmark, France, South Africa (2), Australia (3), New Zealand, Korea and Japan. Additional programs in England and Spain (2) hold accreditation candidate status. Several other chiropractic programs around the world are pursuing accreditation, and some are new or under new development including Mexico (2), Brazil (2), Switzerland and Malaysia.

Curriculum content
The chiropractic curriculum typically includes courses in:

- Anatomy
- Biochemistry
- Physiology
- Microbiology and Immunology
- Pathology
- Public Health
- Clinical Skills (including history and physical examination)

- Clinical and Laboratory Diagnosis
- Clinical Sciences (including the study of cardiopulmonary, gastrointestinal and genitourinary disorders; dermatology; ophthalmology; otolaryngology)
- Gynecology and Obstetrics
- Pediatrics
- Geriatrics
- Diagnostic Imaging (procedures and interpretation)
- Psychology and Abnormal Psychology
- Nutrition and Clinical Nutrition
- Biomechanics
- Orthopedics
- Neurology
- Emergency Procedures and First-Aid
- Spinal Analysis
- Principles and Practice of Chiropractic
- Clinical Reasoning and Decision Making
- Chiropractic Manual Therapy and Adjustive Procedures
- Research Methods and Statistics
- Professional Practice Ethics and Office Management

There are many opportunities for postgraduate study in chiropractic. A number of full-time residency programs exist, of which the most popular and ubiquitous is diagnostic imaging (a three year, full-time residency). A full-time residency program in chiropractic geriatrics was recently initiated at Northwestern Health Sciences University, and National University of Health Sciences offers three-year residency programs in family practice and research.

Other institutions are offering additional degree programs. Numerous certification programs exist in a variety of subject areas such as orthopedics, pediatrics, sports injuries, and nutrition, and are typically taught at chiropractic colleges or through professional associations. Finally, a growing number of institutions are offering accredited master's degrees in chiropractic-related fields. These degree programs are completed following either part-time (including hybrid and online learning) or full-time residential programs at institutions that offer the Doctor of Chiropractic degree program. Examples are Master of Science degrees in Nutrition/Functional Medicine and Sports and Exercise Science at the University of Western States, Applied Clinical Nutrition at New York Chiropractic College, Health Promotion at Cleveland Chiropractic College, Advanced Clinical Practice at National University of Health Sciences, Sports science and rehabilitation at Logan College of Chiropractic, and Master of Health Science degrees in Clinical Nutrition and Clinical Chiropractic Orthopedics at Northwestern Health Sciences University.

Accreditation

Each institution offering a Doctor of Chiropractic degree program in the United States is a member of the Association of Chiropractic Colleges (ACC) and is accredited by the Council on Chiropractic Education (CCE). CCE is the agency recognized by the United States Secretary of Education for accreditation of programs and institutions offering the Doctor of Chiropractic degree. CCE ensures the quality of chiropractic medical education in the US by means of accreditation, advancing educational improvement, and providing public information. CCE develops accreditation criteria to assess how effectively programs or institutions plan, implement and evaluate their mission and goals, program objectives, inputs, resources, and outcomes of their chiropractic programs.

The CCE is also recognized by the Council for Higher Education Accreditation (CHEA) and is a member of the Association of Specialized and Professional Accreditors (ASPA).[5] All but one of the institutions offering chiropractic programs in the United States also maintain institutional accreditation through regional post-secondary accrediting associations.

Regional chiropractic accrediting bodies are also located in Canada (www.chirofed.ca), Europe (www.cce-europe.org), and Australasia (www.ccea.com.au). Together, these accrediting bodies have formed the Councils on Chiropractic Education International (CCEI). CCEI is committed to excellence in chiropractic education through its *International Chiropractic Accreditation Standards* and through aid in the development and recognition of new accrediting bodies in geographic regions where such agencies are not currently established.[6]

CCEI member agencies provide accreditation services to chiropractic educational entities throughout the world. Accreditation decisions and status designations by CCEI member agencies are mutually endorsed on the basis of membership in CCEI. CCEI has established the following goals:

1. Define minimal model educational standards and ensure their adoption and maintenance by accrediting agencies worldwide;
2. define the process of accreditation and assure appropriate implementation and administration of the process by accrediting agencies worldwide;
3. establish and maintain a process for verifying the equivalence of the educational standards and accreditation process utilized by CCEI member accrediting organizations worldwide;
4. assist and provide guidance for the development of accrediting agencies toward their full autonomy and membership in CCEI;
5. promote a continuous model of educational standards, recognizing educational, cultural and legislative diversity in various countries and regions; and

6. advocate quality education through the dissemination and promotion of information to governments, professional organizations, and others.[7]

Regulation and Certification

Regulatory status

Doctors of Chiropractic are regulated in all jurisdictions of the United States and Canada. Licensing boards from each state, and from many other jurisdictions outside the US, are members of the Federation of Chiropractic Licensing Boards (FCLB). The FCLB mission statement is "To protect the public and to serve our member boards by promoting excellence in chiropractic regulation."[8]

Licensing laws differ among the various jurisdictions. Some states such as Oregon, Illinois, Oklahoma, and New Mexico allow a broader scope of practice, while other states such as Washington and Michigan are more restrictive. Efforts are under way to develop and expand the scope of practice in most jurisdictions, to allow for the integration of contemporary practices and to better serve patients that seek care from chiropractic physicians. Members of jurisdictional licensing boards are generally governmental appointees and serve for established lengths of terms. Most boards also have public members that serve alongside the members of the chiropractic profession. More information on scope of practice can be found on the FCLB website. The FCLB also maintains a listing of actions taken against individual chiropractors in their Chiropractic Information Network-Board Action Databank (CIN-BAD), accessible on their website for a fee.

Similar organizations operate in Canada, Europe, and Australia. Local jurisdiction regulations also vary in each of these countries. Chiropractic medicine is also practiced in many countries where official legal recognition has not yet occurred; in unregulated locations, those claiming to practice chiropractic often come from very diverse backgrounds and may lack any formal education or training beyond a few seminars. The World Federation of Chiropractic has been operational for 20 years and has had significant influence and success in promoting legislation and standards for chiropractic practice.

Examinations and certifications

The National Board of Chiropractic Examiners (NBCE) is the principal testing agency for the chiropractic profession. Established in 1963, NBCE develops and administers standardized national examinations according to established guidelines. NBCE is dedicated to promoting excellence in the chiropractic profession by providing testing programs that measure educational attainment and clinical competency of those seeking licensure. Their examinations serve the needs of state licensing authorities, chiropractic colleges, educators and students, doctors of chiropractic, and the public. These examinations serve the profession and public by:

- promoting high standards of competence
- assisting state licensing agencies in assessing competence
- facilitating the licensure of newly graduated chiropractors, thereby enhancing professional credibility

In providing standardized written and performance assessments for the chiropractic profession, the NBCE develops, administers, analyzes, scores, and reports results from various examinations. The NBCE scores are among the criteria utilized by state licensing agencies to determine whether applicants demonstrate competency and satisfy state qualifications for licensure.

In its expanding role as an international testing agency, the NBCE espouses no particular chiropractic philosophy, but formulates test plans according to information provided collectively by the chiropractic colleges, the state licensing agencies, field practitioners, subject specialists, and a *Job Analysis of Chiropractic*.[9] There are four national exams students take in order to apply for licensure (most states accept the NBCE exam as their basic examination for licensure). The Part IV Exam is a practical clinical exam designed to test for competency and has gained international attention from other disciplines.[10] Recently, an International Board of Chiropractic Examiners was formed for the purpose of standardizing the formal testing of chiropractic candidates for licensure outside the US.

Research

For 60 years after its founding in 1946, the Foundation for Chiropractic Education and Research was the central focus of research in the profession. With the significant growth of research capacity in the chiropractic medical schools and the increasing availability of government grants, the center of research action in chiropractic medicine has shifted to these academic centers. Researchers on chiropractic with Palmer Chiropractic College, University of Western States, Northwestern Health Sciences University, Southern California University of Health Sciences, National University of Health Sciences, and New York Chiropractic College are among those in the United States who have received significant federal grants.

Palmer Chiropractic College has operated an NIH consortium center program for the last ten years, and has been involved with organizing the chiropractic profession's Research Agenda Conference (RAC). This meeting has been partially supported with federal grants. The Association of Chiropractic Colleges (ACC) sponsors annual conferences in which educators from around the world meet and present research papers and posters. Since the 1990s, this annual meeting has been combined with the

RAC. These joint ACC-RAC conferences are attended by the academic and research community of chiropractic, as well as chiropractic practitioners, basic scientists and researchers from related medical and health care professions. Roughly $30 million in NIH grants was awarded to chiropractic colleges and universities from 1999-2012.

A significant portion of industry support for the profession has also been funneled into research, including over $10 million from the National Chiropractic Mutual Insurance Company (NCMIC). Funds from Foot Levelers, Inc. have supported numerous fellowships, helping Doctors of Chiropractic obtain graduate research training at the Master's and PhD levels. Through such fellowships over the past 40 years, more than 100 Doctors of Chiropractic have earned graduate degrees (Master and Doctorate) in academic disciplines at major universities.

For the last ten years, the World Federation of Chiropractic (WFC) has played an important role in sponsoring biennial international research symposia, both in the US and internationally. These events have provided venues for research and educational specialists on an international level to collaboratively present and discuss new developments in chiropractic health care and education. The WFC provides a forum for dialogue and exchange on a world-wide basis and has been productive in establishing common ground for the profession in individual countries around the world as well as with the World Health Organization.

At the federal level in the US, within the structure of the National Institutes of Health (NIH), the formation of what started as the Office for Alternative Medicine has now grown to become the National Center for Complementary and Alternative Medicine (NCCAM). The change from having no presence at the NIH, to the formation of an Office with a budget of $2 million, to the formation of a Center with a budget in excess of $100 million is indeed evidence of dramatic growth. The chiropractic profession has played a significant role in the growth and recognition of complementary and alternative medicine through its research, educational, and legislative initiatives. A Doctor of Chiropractic was the first member of a complementary and alternative profession to be hired as program officer by the NIH.

Scientific journals
The ACC publishes the peer reviewed *Journal of Chiropractic Education*, which is indexed in PubMed. National University of Health Sciences (NUHS), through Elsevier Publishing, supports the publication of three peer-reviewed and indexed journals: *Journal of Manipulative and Physiological Therapeutics* (JMPT), *Journal of Chiropractic Medicine* (JCM), and an online journal, *Journal of Chiropractic Humanities* (JCH). NUHS has published JMPT since 1978.

Challenges and Opportunities

Key challenges 2013 – 2018

- further expansion in federal healthcare delivery systems; healthcare reform; and agreements between chiropractic educational institutions and Veteran Administration facilities
- developing programs to improve enrollment trends at chiropractic educational institutions
- providing funding for research programs, residencies, and fellowships at all chiropractic institutions
- implementing measure to address the high cost of education
- exploring measures to provide fair compensation for Doctors of Chiropractic in clinical practice
- developing more opportunities for integration and collaboration with other healthcare professions
- better informing the public about the benefits of chiropractic health care

- achieving political unity within the profession
- adapting the practice of chiropractic to the realities of practicing evidence-based care in a new healthcare reform environment, where cost, outcomes, and accountability are key elements

Key opportunities 2013-2018

- changes to the US healthcare delivery system with the inclusion and integration of chiropractic medical services on a broader scale
- continued growth and acceptance of complementary and integrative health care by the public
- ongoing and increased research support through federal agencies such as NIH
- continued development of chiropractic education and integration within larger university systems
- more collaborative work among the CAM professions
- continued growth and expansion of the chiropractic profession on a worldwide basis

Resources

Organizations and websites

In the US, there are two national organizations representing chiropractors: the American Chiropractic Association and the International Chiropractors Association. The World Federation of Chiropractic represents the profession on a global basis. There are approximately 88,000 chiropractic practitioners worldwide.

- American Chiropractic Association
  www.acatoday.org
- Association of Chiropractic Colleges
  www.chirocolleges.org

- Council on Chiropractic Education
  www.cce-usa.org
- Councils on Chiropractic Education International
  www.cceintl.org
- Federation of Chiropractic Licensing Boards
  www.fclb.org
- International Chiropractors Association
  www.chiropractic.org
- National Board of Chiropractic Examiners
  www.nbce.org
- World Federation of Chiropractic
  www.wfc.org

Bibliography

Christensen MG, Kollasch MW. *Job Analysis of Chiropractic 2005*. Greeley, CO: National Board of Chiropractic Examiners; 2005.

Haldeman S, ed in chief. *Principles and Practice of Chiropractic*. 2nd ed. New York, NY: McGraw-Hill; 2004.

Leach RA. *The Chiropractic Theories: A Textbook of Scientific Research*. 4th ed. Baltimore, MD: Lippincott Williams & Wilkins; 2004.

Peterson DH, Bergmann TF. *Chiropractic Technique*. 2nd ed. St. Louis, MO: Mosby; 2000.

Phillips, RB. The chiropractic paradigm. *J Chiropr Educ*. 2001; 15(2):49-52.

Citations

1. Council on Chiropractic Education in the United States. Accreditation Standards – Principles, Processes & Requirements for Accreditation. January 2012. http://www.cce-usa.org/Publications.html
2. World Health Organization. WHO Guidelines on basic training and safety in chiropractic. World Health Organization. Geneva 2005.

http://whqlibdoc.who.int/publications/2006/9241593717_eng.
pdf

3. Heyboer M. Your piece of the pie. *Chiropractic Economics.*
   June 2011; 8: 32-46.
   http://www.chiroeco.com/joomla/images/stories/2011SalaryS
   urvey.pdf

4. Heyboer M. The big picture. *Chiropractic Economics.* May
   2012; 8: 42-62.
   www.chiroco.com/joomla/images/stories/SE_Survey2012.pdf

5. Council on Chiropractic Education in the United States.
   www.cce-usa.org

6. Councils on Chiropractic Education International.
   International Chiropractic Accreditation standards.
   http://www.cceintl.org/uploads/2010-04-
   26_CCEI_International_Chiropractic_Accreditation_Standar
   ds_vfd_5_09.pdf

7. Councils on Chiropractic Education International.
   http://www.cceintl.org

8. Federation of Chiropractic Licensing Boards. www.fclb.org

9. National Board of Chiropractic Examiners. *The 2005 Job
   Analysis.* National Board of Chiropractic Examiners. Greeley,
   Colorado, 2005.

10. National Board of Chiropractic Examiners. www.nbce.org

Massage Therapy

First Edition (2009) Authors: Jan Schwartz, MA;
Cherie Monterastelli, RN, MS, LMT

Second Edition (2013) Editors: Jan Schwartz, MA;
Stan Dawson, DC, NMT, CMT; Pete Whitridge, BA, LMT

First Edition (2009) Partner Organization:
American Massage Therapy Association –Council of Schools
(AMTA-COS)

Second Edition (2013) Partner Organization:
Alliance for Massage Therapy Education (AFMTE)

About the Authors/Editors: Schwartz is past Chair of the Commission on Massage Therapy Accreditation, co-owner of Education and Training Solutions, LLC, is a member of the ACCAHC Board and Executive Committee and is Co-chair of the ACCAHC Education Working Group. Monterastelli previously served as a board member of the American Massage Therapy Association – Council of Schools and ACCAHC. Dawson is an ACCAHC Board member, a member of the ACCAHC Education Working Group, Vice President of Alliance for Massage Therapy Education, and owner and Executive Director of ASHA, School of Massage. Whitridge is the President of Alliance for Massage Therapy Education, former Chair of the Florida Board of Massage Therapy, and the Accreditation Compliance Officer for the Florida School of Massage.

Philosophy, Mission, Goals

Several organizations and scholars have created definitions that incorporate the mission and philosophy of massage therapy (MT). For example, in the frequently used textbook, *Tappan's*

Handbook of Healing Massage Techniques, the definition is: "Massage is the intentional and systematic manipulation of the soft tissues of the body to enhance health and healing." [1(p4)] Similarly, according to the American Massage Therapy Association (AMTA) Glossary of Terms, massage therapy is "a profession…with the intention of positively affecting the health and well-being of the client through a variety of touch techniques."

Because most definitions of massage therapy include the use of touch, which is a basic, non-technological approach to health and healing, it follows that most massage therapists subscribe to a natural healing philosophy. That philosophy encompasses a preference for "natural methods of healing, the belief in an innate healing force, and a holistic view of human life." [1(p14)]

With the advent of the Patient Protection and Affordable Care Act of 2010, the healthcare system shows signs of shifting toward collaboration between conventional and complementary healthcare professions, with the goal of wellness in addition to treating illness, and physicians' pay based on health outcomes. With doctor behavior incentivized by payment for keeping people healthy rather than performing more tests and procedures, massage can be particularly well suited to this shift in philosophy. The focus of most massage therapists on health improvement, a positive patient experience combined with the relatively low cost of massage treatment are aligned with the concerns of the Triple Aim framework developed by the Institute of Healthcare Improvement.

History of the profession

Massage has been practiced in most cultures in both the East and the West throughout human history. Traditional peoples used a variety of techniques that are now known as massage, all of which included some form of person-to-person touching with the intention of manipulating and relaxing the muscles of the body. From the South Sea Islands to the Mexican peninsula and the indigenous cultures of the Americas to the ancient civilizations of Greece, Asia and Africa, historians and researchers find evidence of the practice of massage. In India, the ancient practice of Ayurveda included forms of movement therapy and massage. In Greece, Hippocrates wrote about the ability of massage to build muscle as well as heal it. China and Japan each developed varieties of natural healing that included touch.

The more modern practice of massage, known as Western massage or Swedish massage, became prominent in the 19th and early 20th centuries. Called the father of Swedish massage, Per Henrik Ling (1776-1839) developed a series of exercises that became known as "medical gymnastics" and used a series of movements that applied resistance to the joints. These techniques, however, have little resemblance to massage as it is known today. While there was a hands-on relationship between therapist and subject, the activities were more like those used by physical therapists than massage therapists. Ling's perspective on the practices he was developing evolved, however, and he began to consider the relationship between the physical and mental aspects of wellness, the mind/body connection, which is today very much a part of Swedish massage and other massage modalities.

Not quite a century later, a Dutch physician named Johann Georg Mezger (1838-1909) developed the techniques that are now the basis for Swedish massage:

- Effleurage: Long, gliding strokes
- Petrissage: Lifting and kneading the muscles
- Friction: Firm, deep, circular rubbing movements
- Tapotement: Brisk tapping or percussive movements

- Vibration: Rapidly shaking or vibrating specific muscles

Researchers noticed the similarities between the work of Ling and Mezger and gave Sweden the credit for developing these techniques.

Massage technique has evolved considerably in the past fifty years as therapists strive to be more helpful to their clients and as massage research has advanced. Research and technique advancement is focused primarily on the effects of massage and bodywork on circulation, lymph fluid circulation, muscle tone, fascial patterns, joint neurology, neuroendocrine effects, craniosacral fluid flow and stages of rehabilitation. Pain research projects and the effects of massage on depression and immune function show considerable promise.

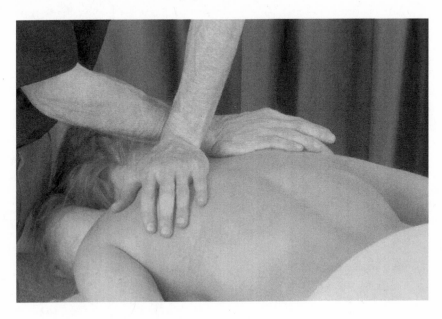

Characteristics and Data[*]

Industry research estimates that there are 280,000-320,000 practicing massage therapists and massage school students in the US. Today's massage therapists are likely to enter the massage therapy profession as a second career and are predominantly female (85%). The average massage therapist is in her early 40s.

Massage therapists earn a comparable annual income to other healthcare support workers, according to the US Department of Labor Statistics. In 2010, the average annual income for a massage therapist was $34,900. A 2012 AMTA survey indicates that the annual income for a massage therapist who provides 15 hours of massage per week is $21,028. More than half of massage therapists (53%) also earned income working in another profession. The majority of massage therapists are self-employed. They work an average of 15 hours a week providing massage and give an average of 41 massages per month.

Growth in the healthcare industry is providing numerous jobs for massage therapists. There has been an increase in the number of massage therapists directly employed by spas and clinics or employed or contracted to work in healthcare settings. From 2005 to 2011, the percentage of massage therapists who worked in a healthcare environment increased from 10% to 22%. According to 2011 survey results released by Health Forum, a subsidiary of the American Hospital Association (AHA) and Samueli Institute, more than 42 percent of responding hospitals indicated they offer one or more complementary and alternativemedicine (CAM) therapies, up from 37 percent in

---

[*] Data from the American Massage Therapy Association Industry Fact Sheet, February, 2012, unless otherwise noted. Detailed information available at amtamassage.org/MTIndustryFactSheet.

2007. Massage therapy is in the top two services provided in both outpatient and inpatient settings.

Clinical Care

Approach to patient care

Massage therapists in clinical settings typically perform an intake interview focusing on the client's health history, chief complaints and main reason for seeking care. After the intake, a massage therapist usually performs an assessment in order to determine contraindications, need for referral, appropriateness of treating and a plan for a treatment session.

During the treatment session, the massage therapist employs a variety of modalities intended to influence blood flow, lymph flow, muscle tone, fascial length and organization, joint dynamics, joint neurology, and/or craniosacral fluid flow dynamics. Treatments in a clinical setting are typically structured into a treatment plan either by the massage therapist or the health professional the massage therapists' work is supporting or complementing.

Swedish massage (Western Massage) is designed to facilitate relaxation, stress relief through parasympathetic dominance. Advanced level massage can support pain and symptom relief, structural/postural balancing and wellness-oriented collaboration with the entire healthcare team. Advanced level massage therapists may also have specialized training in lymphatic drainage, myofascial manipulation, neuromuscular therapeutic techniques, visceral manipulation, infant massage, pregnancy massage, oncology massage, geriatric massage, sports massage, Asian bodywork therapies or other specialties in massage and bodywork.

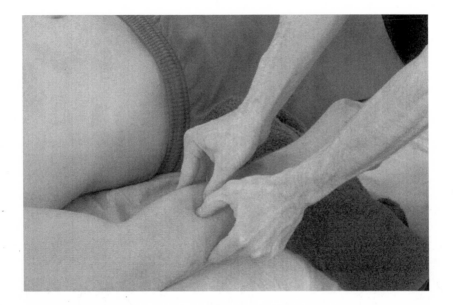

Scope of practice

The scope of practice and common terms are similar in most of the state regulations, but there are significant differences between states. All describe massage as soft tissue manipulation to improve health or some similar idea. All exclude the authority to make a diagnosis, prescribe medications, perform chiropractic adjustments, offer injections or venipuncture, or practice acupuncture, or psychotherapy. Entry level training prepares practitioners for a part of the potential scope of practice afforded massage therapists. Practitioners are *ethically* obligated to limit their practice to what they have been adequately trained to do.

Referral practices

Most massage therapists are adequately trained to know how and when to refer to other healthcare professionals. Because of inconsistencies in training and experience, and the broad range of educational requirements noted below, practitioners' skills in the area of referral practices can vary. Advanced practitioners, however, appreciating the interdependence of mental, emotional, musculoskeletal and visceral elements of the spirit-mind-body system, value collaboration with both conventional and other complementary healthcare professionals. Mature professionals understand the appropriateness of helping their clients find care that is efficacious, cost effective and patient-centered. This leads to an attitude of comfort with referral to the appropriate healthcare practitioner whenever indicated.

Since massage therapists do not diagnose, they need to confer on cases with physicians and other healthcare practitioners in order to judge case management issues, when indicated. Massage therapists are trained to refer to lists of possible contraindications for which referral may be required or advised.

Third-party payers

Some third parties reimburse for massage therapy, however, massage therapy is not covered in Centers for Medicare and Medicaid Services legislation. Payments are most often made for workers' compensation claims and auto accident injuries. Additional reimbursements vary by state and insurance provider. The state of Washington, for instance, mandates that all health plans include every category of licensed provider in their benefit design. Massage therapists are receiving an increasing number of referrals for healthcare treatments.

The AMTA holds a seat on the American Medical Association's Health Care Professionals Advisory Committee (CPT-HCPAC). This assures representation of the profession in discussions of coding issues that may affect massage therapy. Issues involving insurance, such as the AMA CPT codes, are forwarded to the AMTA board of directors for consideration of what path would be in the best interest of the massage profession as a whole.

Integration Activities

A number of partnerships have developed between massage schools and medical schools. These partnerships can be as minimal as medical schools referring students to massage schools in the local area or as involved as jointly developing a multi-disciplinary student clinic or a curriculum for use in each other's institutions.

Massage therapists are working more and more in integrated settings like hospitals, wellness centers, chiropractic offices, and settings such as medical, physical therapy and sports medicine, acupuncture and Oriental medicine settings, and holistic health centers.

Massage therapists working with ACCAHC have helped in the development of the ACCAHC Competencies for Optimal Practice in Integrated Environments which the Alliance for Massage Therapy Education has endorsed. Massage therapists working with ACCAHC have also been involved in interprofessional education (IPE) conferences and research conferences. If the massage profession adopts the competencies and prepares massage therapists in these competencies, massage will be positioned to enter the healthcare system of the future with its focus on integrated care.

Education

Schools and programs

According to the Associated Bodywork and Massage Professionals (ABMP) 2011 enrollment survey, there are approximately 1,400 state-approved massage therapy training programs in the US. Of those schools/programs, 57% have been accredited by agencies recognized by the US Department of Education. While the number of massage therapy training institutions has grown dramatically since the late 1990s, the number of training programs has declined slightly in the past five years.

Of these massage school programs in the US, nearly 74% have programs of 500 hours or more with the average at 650 hours. In most cases, the total number of hours a school offers trends closely to what is required by that state for licensure. This varies from state to state. The minimum number of hours required by the Federal Department of Education for national accreditation for massage programs is 600.

The Alliance for Massage Therapy Education was organized in 2009 as a 501c 6 non-profit corporation. It was created to "fill a gap" as the profession was missing a not-for-profit dedicated to education. There are three member groups represented by the Alliance: schools, continuing education providers, and teachers. The AFMTE was designated by the ACCAHC board of directors to represent massage educators in October 2011. The AFMTE is dedicated to improving the quality of massage therapy education by raising teacher standards, curriculum standards and education standards. The AFMTE created a set of core competencies for massage therapy teachers in 2012 to be used as the basis for a national teacher certification program and for implementation in accreditation standards and state regulations to bolster the quality of massage education. They also hold an annual conference on massage education.

A national membership organization, a national business with a membership component, and several state organizations provide services and support to schools. The two national groups are the not-for-profit American Massage Therapy Association (AMTA) and the for profit Associated Bodywork and Massage Professionals (ABMP).

AMTA offers school memberships that includes benefits such as an annual leadership conference, a teacher education track at the annual convention, webinars, and monthly interactive conference calls known as "Teacher Talk" for faculty, an education calendar for schools to list their continuing education programs, and optional school liability and business property insurance. AMTA has long been an advocate of fair and consistent legislation regarding massage therapy. The Government Relations Committee works with schools and chapters to lobby and develop regulations that support high standards for the profession. The organization also offers an annual AMTA Schools Summit.

The ABMP offers a complimentary membership to AFMTE with a paid school membership in ABMP. This organization has a staff of six professionals who work directly with the schools. School memberships include materials and resources for students and faculty, classified ads and job postings, an e-mail account and online networking forums, optional school liability and business property insurance. Most importantly, the ABMP offers an annual ABMP School Issues Forum for schools to network and discuss common challenges. The ABMP has an extensive website with archived webinars, teaching tips, learning checklists, and rubrics to support faculty and student success. There is also a regional program known as *Teachers on the Frontline*. These are free one-day seminars to support professional teacher development.

Curriculum content

Unlike the other disciplines within ACCAHC, there are neither national standards nor a single recognized body that the whole massage profession has designated to determine the knowledge and skills required to call one a member of the field. Although there is a specialized accrediting agency for massage therapy which is recognized by the US Department of Education, the Commission on Massage Therapy Accreditation (COMTA), which maintains curriculum competencies, fewer than 10% of massage programs are currently accredited by this agency. There exists, therefore, a wide variety in the type and quality of education available in massage schools and programs. At this time, there is no consensus regarding the required number of hours of study for entry-level massage. The majority of states are at 500 hours and this is also the necessary minimum to sit for the national certification exams. For schools that are not COMTA-accredited, the national certification exams have been the most consistent requirement for many years. Consequently, many schools shaped their curriculum around the required content areas for those exams. The National Certification Board for Therapeutic Massage and Bodywork (NCBTMB) has a process for authorizing applicants to take the exams based on completion of training in these core areas:

- 125 hours of instruction in the body's systems and anatomy, physiology and kinesiology
- 200 hours of in-class, supervised hands-on instruction in massage and bodywork assessment, theory and application instruction
- 40 hours of pathology
- 10 hours of business and ethics instruction (a minimum of 6 hours in ethics)
- 125 hours of instruction in an area or related field that theoretically completes your massage program of study

Many practitioners choose to take continuing education courses to gain more extensive training in a specific modality or area of practice. Requirements for continuing education vary from state to state depending on state licensure requirements, according to the membership organization a practitioner chooses to join, and depending on whether the practitioner elects to maintain the national certification credential.

Until national curriculum standards are adopted by the profession, state regulations, certification and accreditation standards dictate curriculum content to the schools. This arrangement is problematical for the maturation of the profession and for public health clarity because the state curriculum requirements are neither rigorous nor consistent, and almost 50% of the schools are not subject to review by any accreditation agency, yet all are allowed to register as "approved" schools.

Accreditation

The Commission on Massage Therapy Accreditation (COMTA) is recognized by the US Department of Education as specializing in massage. COMTA was started by the massage profession in 1982 and granted recognition by the US Department of Education in 2002. COMTA provides either institutional or programmatic accreditation. Both designations include a comprehensive evaluation of curriculum competencies. Institutional accreditation is available to schools teaching only massage, bodywork and/or aesthetics. Programmatic accreditation is available to massage therapy and bodywork programs within schools which already have institutional accreditation from another recognized agency.

Accreditation by COMTA requires that at least one program in massage or bodywork offered by the school contains 600 clock hours of content and includes teaching and assessment of the required competencies. The minimum clock hour requirement was increased from 500 to 600 in 2001 with a goal of eventually increasing the amount again when enough schools were prepared to offer longer training. In addition to the content areas required by NCBMTB exams, the COMTA competencies outline requirements in areas such as medical terminology, research literacy, client communication and professional relationships.

Regulation and Certification

Regulatory status

Forty-four states and the District of Columbia have some form of regulation in place for massage therapy. There is a lack of consistency among these licensure laws. Depending on the state law, massage therapists can be referred to as licensed, state certified, or registered. In most cases, only individuals who have the state designation may perform massage or use a title indicating that they perform massage.

Examinations and certifications

The National Certification Board for Therapeutic Massage and Bodywork (NCBTMB) is an independent, private, nonprofit organization that was founded in 1992 to establish a certification program and uphold a national standard of excellence. NCBTMB's exam programs are accredited by the Institute for Credentialing Excellence (ICE). There are currently more than 75,000 nationally certified massage practitioners, located in every state in the US and the District of Columbia.

Beginning in 2013, NCBTMB's current national certification will transition to a board certification in massage therapy. This credential will require more than the minimum education required by most state licensing requirements, a predetermined number of hours of documented hands-on work experience, a waiting period of a minimum of 6 months from graduation plus passage of NCBTMB's exam. The exam is designed to measure critical thinking rather than simple recall. Additionally, the program includes adherence to the NCBTMB Code of Ethics and Standards of Practice, current CPR certification, passing a national background check along with other eligibility criteria. The last component of this credential is a commitment to lifelong learning and re-certification. This new credential will be the first of its kind for the massage profession, offering a true separation between licensing and certification.

NCBTMB entry-level licensing exams (NCETM and NCETMB) are used in 38 states and the District of Columbia in either statute or rule. To keep with best past practices, NCBTMB conducts a Job Task Analysis (JTA) every five years to ensure content relevancy. NCBTMB conducted its fifth JTA in 2012 for the licensing exams.

The Federation of State Massage Therapy Boards (FSMTB), established in 2005, is currently made up of 32 state licensing boards and agencies that regulate the massage therapy and bodywork profession. The mission of the FSMTB is to support its member boards in their work to ensure that the practice of massage therapy is provided to the public in a safe and effective manner. In October 2007, FSMTB created an entry-level national licensure examination to serve the needs of the regulatory community in licensing massage therapists after completing a JTA. The FSMTB concluded a second JTA in 2012. Currently, 40 states accept the FSMTB exam. For detailed results and a state-by-state compendium of laws, please visit the FSMTB website listed in the Resources section.

Research

The Massage Therapy Foundation's (MTF) mission is to advance the knowledge and practice of massage therapy by supporting scientific research, education, and community service. The Foundation funds research studies investigating the many beneficial applications of massage therapy. Foundation research grants are awarded to individuals or teams conducting studies that promise to advance the understanding of specific therapeutic applications of massage, public perceptions of and attitudes toward massage therapy, and the role of massage therapy in healthcare delivery.

The Massage Therapy Foundation commissioned a research agenda in 1999, recommending areas of research in massage therapy and bodywork that are most needed. This agenda will be reviewed in 2013. Investigators who apply for MTF research grants are referred to the agenda and are encouraged to address one of the following in their research:

- Build a massage research infrastructure
- Fund studies into safety and efficacy
- Fund studies of physiological (or other) mechanisms (how massage works)
- Fund studies stemming from a wellness paradigm
- Fund studies into the profession of therapeutic massage and bodywork

The MTF launched the *International Journal of Therapeutic Massage and Bodywork: Research, Education and Practice* (IJTMB) in August, 2008. It is a free, online, peer-reviewed journal.

In addition to its efficacy for muscle and other soft tissue ailments, research has shown that massage therapy can relieve symptoms associated with many health issues, such as osteorarthritis.[2] Among the other ways massage therapy has been shown to be effective are:

- **Relief of back pain**
  More than 100 million Americans suffer from low-back pain, and nearly $25 billion a year is spent in search of relief. A 2003 study showed that massage therapy produced better results and reduced the need for painkillers by 36% when compared to other therapies. Today, massage therapy is one of the most common ways people ease back pain.[3]
- **Treating Migraines**
  Of the 45 million Americans who suffer from chronic headaches, more than 60% suffer from migraines. For many,

it's a distressing disorder that is triggered by stress and poor sleep. In a recent study, massage therapy recipients exhibited fewer migraines and better sleep quality during the weeks they received massage, and the three weeks following, than did participants who did not receive massage therapy. Another study found that in adults with migraine headaches, massage therapy decreased the occurrence of headaches, sleep disturbances, and distress symptoms. It also increased serotonin levels, believed to play an important role in the regulation of mood, sleep, and appetite.[4,5]

- **Easing Symptoms of Carpal Tunnel**
  Carpal tunnel syndrome is a progressively painful condition that causes numbness and tingling in the thumb and middle fingers. Traditional treatments for carpal tunnel range from a wrist brace to surgery. However, a 2004 study found that carpal tunnel patients receiving massage reported significantly less pain, fewer symptoms, and improved grip strength compared to patients who did not receive massage.[6]

- **Reducing Anxiety**
  An estimated 20 million Americans suffer from depression. A review of more than a dozen massage studies concluded that massage therapy helps relieve depression and anxiety by affecting the body's biochemistry. In the studies reviewed, researchers measured the stress hormone cortisol in participants before and immediately after massage and found that the therapy lowered levels by up to 53%. Massage also increased serotonin and dopamine, and neurotransmitters that help reduce depression.[7]

- **Alleviating Symptoms and Side Effects of Cancer**
  Massage therapy is increasingly being applied to symptoms experienced by cancer patients, such as nausea, pain, and fatigue. Researchers at Memorial Sloan-Kettering Cancer

Center asked patients to report the severity of their symptoms before and after receiving massage therapy. Patients reported reduced levels of anxiety, pain, fatigue, depression, and nausea, even up to two days later.[8] In a study of breast cancer patients, researchers found that those who were massaged three times a week reported lower levels of depression, anxiety and anger, while increasing natural killer cells and lymphocytes that help to battle cancerous tumors.[9,10]

- **Lowering Blood Pressure**
  Hypertension, if left unchecked, can lead to organ damage. Preliminary research shows that hypertensive patients who received three 10-minute back massages a week had a reduction in blood pressure, compared to patients who simply relaxed without a massage.[11]

Challenges and Opportunities

The following descriptions are intended to assist the reader in understanding aspects of the massage therapy profession rather than formal positions taken by the massage field on the issues.

Key challenges 2012-2015

- There is inconsistent use of terminology and definitions within the massage therapy profession. To continue to move toward consistency, terminology will need to be standardized. Revising and adopting the Massage Therapy Body of Knowledge (MTBOK) would be a strong step toward accomplishing this.
- Massage therapy educational programs need to move to greater standardization in the content, scope, and length of entry-level training programs. This could be accomplished

by revising educational standards based on MTBOK, COMTA and the Entry Level Analysis Project (ELAP).

- Faculty qualifications vary greatly from school to school with no commonly agreed upon formal training for massage therapy instructors. Such training is also part of creating consistency in educational programs for massage therapy. Without consistency in educational programs and faculty qualifications, consistency in terminology will prove difficult. Consumer expectations will also be affected. Adopting the AFMTE Teacher Standards and basing teacher training on them could be significant steps toward accomplishing this.

Key opportunities 2012-2015

- Further development of industry partnerships with potential massage therapy employers will increase knowledge of massage therapy training and credentials among those industries, and provide more opportunities for collaboration in client care. Partnership with employers will, in turn, give the massage education institutions much needed information on what employers' needs are so curriculum can be developed to meet those needs and enhance the likelihood of job placement for graduates.

- Development of the profession's research base will allow practitioners to provide more effective, outcome-based massage therapy for their clients, which will provide better results and increase the demand for massage therapy. Establishing awareness of and interest in massage as an effective therapy will increase acceptance by healthcare providers and third-party payers. Greater communication within the profession will also increase as a standardized scientific language is more commonly used.

- As massage therapy becomes a more integral component of the complementary, alternative, and integrative aspects of health care, there will be an increasing demand for quality, standardization, and research. This demand will help set the parameters for the profession to evolve into a more recognized, valuable, and respected part of health care.

- Improvement of industry partnerships with all healthcare professions, both conventional and complementary and alternative.
- Creation of a Model Practice Act to standardize terminology, to create consistency in state statutes and to improve teacher training standards, educational standards, curriculum standards and facilitate portability for licensed practitioners moving state to state.
- Further utilization and promotion of COMTA programmatic accreditation to assure consistent curriculum content, educational quality and faculty qualifications.

Resources

Organizations and websites
The profession is represented by a number of member organizations and one national certification organization:

- Alliance for Massage Therapy Education (AFMTE) was founded in 2009 as a nonprofit organization as an independent voice, advocate, and resource for the quality of massage education. AFMTE is comprised of massage schools, continuing educators in massage, and massage school faculty. AFMTE represents the education aspect of massage therapy (council of colleges/schools) within ACCAHC. The Alliance is working to advance massage education by developing a set of competencies for massage faculty; a training program for massage faculty based on those competencies; a revision of the MTBOK to clarify definitions and terminology for entry level massage education, advanced level massage education, and specialty certifications; curriculum standards; higher education standards as well as support for a Model Practice Act.

- American Massage Therapy Association (AMTA) is a not-for-profit membership organization that represents more than 150 massage therapy schools and programs and has more than 58,000 member massage therapists. AMTA works to establish massage therapy as integral to the maintenance of good health and complementary to other therapeutic processes, and to advance the profession by promoting certification and school accreditation, ethics and standards, continuing education, professional publications, legislative efforts, public education, and fostering the development of members. AMTA offers liability insurance to its members.

- Associated Bodywork and Massage Professionals (ABMP) is a private business with professional membership dimensions founded in 1987 to provide massage and bodywork practitioners with professional services, information, and public and regulatory advocacy. ABMP has initiatives in the areas of promoting ethical practices, protecting the rights of practitioners, and educating the public regarding the benefits of massage and bodywork. Its current membership totals over 82,000. Members must adhere to a published code of ethics.

- Federation of State Massage Therapy Boards (FSMTB) was formed in 2005 to bring the regulatory community together and provide a forum for the exchange of information. The result of the exchange was the development of a licensure exam which was introduced in 2007. The FSMTB's mission is to support its member boards in their work to ensure that the practice of massage therapy is provided to the public in a safe and effective manner. There are currently 41 member boards with 40 states and territories accepting the exam for licensing.

- National Certification Board for Therapeutic Massage and Bodywork (NCBTMB) was founded in 1992 as a nonprofit organization to establish a national certification program and uphold a national standard of professionalism. NCBTMB works to foster high standards of ethical and professional practice through a recognized, credible credentialing program that assures the competency of practitioners of therapeutic massage and bodywork. NCBTMB also has outreach programs for stakeholders, including schools and state boards. The organization had certified more than 90,000 massage therapists and bodyworkers as of the date of publication of this booklet. NCBTMB's certification program is made up of a number of components: eligibility, examinations, code of ethics, standards of practice, continuing education and recertification. NCBTMB's certification program is accredited by the National Commission for Certifying Agencies (NCCA), the accrediting branch of the National Organization for Competency Assurance (NOCA).

- Alliance for Massage Therapy Education (AFMTE)
  www.afmte.org
- American Massage Therapy Association (AMTA)
  www.amtamassage.org
- Associated Bodywork and Massage Professionals (ABMP)
  www.abmp.com
- Commission on Massage Therapy Accreditation (COMTA)
  www.comta.org
- Federation of State Massage Therapy Boards (FSMTB)
  www.fsmtb.org
- Massage Therapy Foundation
  www.massagetherapyfoundation.org
- Massage Therapy Research Consortium
  www.massagetherapyresearchconsortium.com

- National Certification Board for Therapeutic Massage and Bodywork (NCBTMB)
  www.ncbtmb.org

Bibliography

Benjamin BE, Sohnen-Moe C. *The Ethics of Touch*. Tucson, Arizona: Sohnen-Moe Associates; 2003.

Biel A. *Trail Guide to the Body*. 4th ed. Boulder, CO: Books of Discovery; 2010.

Cohen B. *Memmler's the Human Body in Health and Disease* (12th ed.). Philadelphia: Lippincott, Williams & Wilkins; 2008.

Dryden T, Moyer C, eds.. *Massage Therapy: Integrating Research and Practice*. Champaign, IL: Human Kinetics; 2012.

Fritz S. *Fundamentals of Therapeutic Massage* 4th ed. St. Louis, Missouri: Mosby; 2009.

Frye B. *Body Mechanics for Manual Therapists: A Functional Approach to Self-care*. 2nd ed. Stanwood, Washington: Fryetag; 2004.

Hymel GM. *ResearchMmethods for Massage and Holistic Therapies*. St. Louis, Missouri: Mosby; 2006.

Lowe W. *Orthopedic Massage: Theory and Technique*. St. Louis, Missouri: Elsevier Health Services; 2003.

Menard MB. *Making Sense of Research: A Guide to Research Literacy for Complementary Practitioners*. Toronto, Ontario, Canada: Curties-Overzet; 2009.

Rattray F, Ludwig L. *Clinical Massage Therapy: Understanding, Assessing and Treating over 70 Conditions*. Elora, Ontario, Canada: Talus; 2001.

Sohnen-Moe C.*Business Mastery: A Guide for Creating a Fulfilling, Thriving Business and Keeping it Successful*. Tucson, AZ: Sohnen-Moe Associates; 2008.

Thompson DL. *Hands Heal: Communication, Documentation, and Insurance Billing for Manual Therapists*.3rd ed. Philadelphia, PA: Lippincott, Williams & Wilkins; 2006.

Werner R. *A Massage Therapist's Guide to Pathology.* 5th ed. Philadelphia, PA: Lippincott, Williams & Wilkins; 2012.

Williams A, ed. *Teaching Massage: Fundamental Principles in Adult Education for Massage Program Instructors.* Philadelphia, PA: Lippincott, Williams & Wilkins; 2008.

Yates J. *A Physician's Guide to Therapeutic Massage.* Toronto, Ontario, Canada: Curties-Overzet; 2004.

Citations

1. Benjamin P, & Tappan F. *Tappan's Handbook of Healing Massage Techniques: Classic, Holistic, and Emerging Methods* 5th ed. Boston, MA: Pearson Education; 2009.

2. Perlman A, Sabina A, Williams AL, et al. Massage therapy for osteoarthritis of the knee: a randomized trial. Arch Intern Med. 2006;166(22):2533-2538.

3. Cherkin DC, Sherman KJ, Devo RA, et al. A review of the evidence for the effectiveness, safety, and cost of acupuncture, massage therapy, and spinal manipulation for back pain. *Ann Intern Med.* 2003;138(11):898-906.

4. Lawler SP, Cameron LD. A randomized, controlled trial of massage therapy as a treatment for migraine. *Ann Behav Med.* 2006;32(1):50-9.

5. Field T, Hernandez-Reif M, Miguel Diego M, et al. Cortisol decreases and serotonin and dopamine increase following massage therapy. *Int J Neuroscience.* 2005;115(10):1397-1413.

6. Cambron, J., Dexheimer, J., Coe, P., Swenson, R. Sidepeffects of massage therapy: a cross sectional study of 100 clients. *J. Comp Alt Med,* 2007; 13(8), 293-796.

7. Cassileth BR, Vickers AJ. Massage therapy for symptom control: outcome study at a major cancer center. *J Pain Symptom Manage.* 2004;28(3):244-9.

8. Hernandez-Reif M, Ironson G, Field T, et al. Breast cancer patients have improved immune and neuroendocrine

functions following massage therapy. *J Psychosom Res.* 2004;57(1):45-52.

9. Moraska A, Chandler C. Changes in psychological parameters in patients with tension-type headache following massage therapy: A pilot study. *J Man Manip Ther* 2009;17(2):86-94.

10. Moyer CA, Rounds J & Hannum J. A meta analysis of massage therapy research. *Psychol Bull.* 20004:130(1):3-18.

11. Olney CM. The effect of therapeutic back massage in hypertensive persons: a preliminary study. *Biol Res Nurs.* 2005;7(2):98-105.

Direct-entry Midwifery

First Edition (2009) Author and Second Edition (2013) Editor: Jo Anne Myers-Ciecko, MPH

Partner Organization: Midwifery Education Accreditation Council (MEAC)

About the Author/Editor: Myers-Ciecko is Senior Advisor for the Midwifery Education Accreditation Council.

Philosophy, Mission, and Goals

Midwives work in partnership with women to give the necessary support, care, and advice during pregnancy, labor and the postpartum period, to conduct births on the midwife's own responsibility, and to provide care for the newborn and the infant. This care includes preventive measures, the promotion of normal birth, the detection of complications in mother and child, the accessing of medical or other appropriate assistance, and the carrying out of emergency measures. Midwives have an important task in health counseling and education, not only for the woman, but also within the family and community. This work involves prenatal education and preparation for parenthood, and may extend to women's health, sexual or reproductive health, and childcare.[1]

Direct-entry midwifery refers to an educational path that does not require prior nursing training to enter the profession. Certified Professional Midwives (CPMs) are direct-entry midwives who are nationally certified, a credential first awarded in 1994. The National Association of Certified Professional Midwives (NACPM) was founded in 2000 to increase women's access to midwives by supporting the work and practice of

CPMs. NACPM adopted the following *Philosophy and Principles of Practice* in 2004:

1. NACPM Members respect the mystery, sanctity, and potential for growth inherent in the experience of pregnancy and birth.
2. NACPM members understand birth to be a pivotal life event for mother, baby, and family. It is the goal of midwifery care to support and empower the mother and to protect the natural process of birth.
3. Members of NACPM respect the biological integrity of the processes of pregnancy and birth as aspects of a woman's sexuality.
4. Members of NACPM recognize the inseparable and interdependent relationship of the mother-baby pair.
5. NACPM members believe that responsible and ethical midwifery care respects the life of the baby by nurturing and respecting the mother, and, when necessary, counseling and educating her in ways to improve fetal/infant well-being.
6. NACPM members work as autonomous practitioners, recognizing that this autonomy makes possible a true partnership with the women they serve and enables them to bring a broad range of skills to the partnership.
7. NACPM members recognize that decision making involves a synthesis of knowledge, skills, intuition, and judgment.
8. Members of NACPM know that the best research demonstrates that out-of-hospital birth is a safe and rational choice for healthy women, and that the out-of-hospital setting provides optimal opportunity for the empowerment of the mother and the support and protection of the normal process of birth.
9. NACPM members recognize that the mother or baby may on occasion require medical consultation or collaboration.

10. NACPM members recognize that optimal care of women and babies during pregnancy and birth takes place within a network of relationships with other care providers who can provide service outside the scope of midwifery practice when needed.

While NACPM represents CPMs, the Midwives Alliance of North America (MANA), founded in 1982, is a broad-based alliance representing the breadth and diversity of the profession of midwifery in the United States. Members include Certified Professional Midwives as well as Certified Nurse-Midwives (CNMs), state-licensed midwives, and traditional midwives who serve special populations such as the Amish or indigenous communities.

MANA "Core Competencies for Basic Midwifery Practice," revised in 2011, include the following guiding principles:

The midwife provides care according to the following guiding principles of practice:

- Pregnancy and childbearing are natural physiologic life processes.
- Women have within themselves the innate biological wisdom to give birth.
- Physical, emotional, psychosocial and spiritual factors synergistically shape the health of individuals and affect the childbearing process.
- The childbearing experience and birth of a baby are personal, family and community events.
- The woman is the only direct care provider for herself and her unborn baby; thus the most important determinant of a healthy pregnancy is the mother herself.

- The parameters of "normal" vary widely, and each pregnancy, birth and baby is unique.

In consideration thereof:

- Midwives work in partnership with women and their chosen support community throughout the caregiving relationship.
- Midwives respect and support the dignity, rights and responsibilities of the women they serve.
- Midwives are committed to addressing disparities in maternal and child healthcare status and outcomes.
- Midwives work as autonomous practitioners, although they collaborate with other healthcare and social service providers when necessary.
- Midwives work to optimize the well-being of mothers and their developing babies as the foundation of caregiving.
- Midwives recognize the empowerment inherent in the childbearing experience and strive to support women to make informed decisions and take responsibility for their own and their baby's well-being.
- Midwives integrate clinical or hands-on evaluation, theoretical knowledge, intuitive assessment, spiritual awareness and informed consent and refusal as essential components of effective decision making.
- Midwives strive to ensure optimal birth for each woman and baby and provide guidance, education and support to facilitate the spontaneous processes of pregnancy, labor and birth, lactation and mother–baby attachment, using appropriate intervention as needed.
- Midwives value continuity of care throughout the childbearing cycle and strive to maintain such continuity.

- Midwives are committed to sharing their knowledge and experience through such avenues as peer review, preceptorship, mentoring and participation in MANA's statistics collection program. [2]

Characteristics and Data

There are approximately 2,000 Certified Professional Midwives (CPMs). Information on the number of direct-entry midwives who are not CPMs is not available, though estimates range from 500 to 3,000. CPMs are typically self-employed, working in small, community-based practices. There are approximately 11,000 certified nurse-midwives (CNMs) practicing in the US. Most CNMs practice in hospitals, while virtually all CPMs attend births in the client's home or freestanding birth centers. Direct-entry midwives own half of all birth centers in the country.

Clinical Care

Approach to patient care

Midwifery care encompasses the normal childbearing cycle of pregnancy, birth, and postpartum care. NACPM, MANA, and other leading organizations that support direct-entry midwifery endorse the following statement:

"The *Midwives Model of Care*™ is based on the fact that pregnancy and birth are normal life events. The Midwives Model of Care includes:

- monitoring the physical, psychological and social well-being of the mother throughout the childbearing cycle;

- providing the mother with individualized education, counseling, and prenatal care, continuous hands-on assistance during labor and delivery, and postpartum support;
- minimizing technological interventions; and
- identifying and referring women who require obstetrical attention.

The application of this model has been proven to reduce the incidence of birth injury, trauma, and cesarean section".[3-9]

Researchers who interviewed direct-entry midwives early in the development of the home-birth movement found that a wellness orientation, shared responsibility, passive management, holistic care, and individualized care were central to the midwifery approach to care.[10] Exemplary midwifery practice was described by clients, direct-entry midwives, and nurse-midwives in another study that identified critical process-of-care qualities such as supporting the normalcy of birth, respecting the uniqueness of the woman and family, vigilance and attention to detail, and creating a setting that is respectful and reflects the woman's needs [5].

The North American Registry of Midwives (NARM) states that, "CPMs work with women to promote a healthy pregnancy, and provide education to help her make informed decisions about her own care. In partnership with their clients they carefully monitor the progress of the pregnancy, labor, birth, and postpartum period and recommend appropriate management if complications arise, collaborating with other healthcare providers when necessary. The key elements of this education, monitoring, and decision making process are based on evidence-based practice and informed consent. "[11]

Scope of practice

The National Association of Certified Professional Midwives (NACPM) defines the midwives' scope of practice as providing expert care, education, counseling, and support to women and their families throughout the caregiving partnership, including pregnancy, birth, and postpartum. NACPM members provide ongoing care throughout pregnancy and continuous, hands-on care during labor, birth, and the immediate postpartum period. They are trained to recognize abnormal or dangerous conditions needing expert help outside their scope and to consult or refer as necessary.

The North American Registry of Midwives (NARM) recognizes that each midwife is an individual with specific practice protocols that reflect her own style and philosophy, level of experience, and legal status, and that practice guidelines may vary with each midwife. NARM does not set protocols for all CPMs to follow, but requires that they develop their own practice guidelines in written form.

In certain jurisdictions, the midwives' scope of practice includes well-woman care and family planning services. Midwives may also administer certain drugs and devices as specified in state law, including IV fluids, antibiotics, local anesthetic, antihemorrhagics for postpartum use, etc. Midwives typically carry oxygen and resuscitation equipment and are certified in adult and neonatal resuscitation.

Referral practices

Direct-entry midwives care for essentially healthy women expecting normal pregnancy, birth, and postpartum experiences. Midwives consult with and refer to an array of healthcare professionals, social service providers, and others whose services may benefit the woman and her family. These include, but are not limited to, allopathic physicians, naturopathic physicians, acupuncturists, chiropractors, massage therapists, childbirth educators, psychologists, family therapists, doulas, nutritionists, social workers, and food support and housing agencies. In some states, midwives are legally obligated to consult with and/or refer women to obstetricians or family physicians with obstetrical privileges when certain conditions arise. In addition, because the vast majority of direct-entry midwives attend births in their clients' homes or in freestanding birth centers, any complications that require hospitalization for labor and/or birth necessitate transfer of care to an obstetrical care provider with hospital privileges. Most transfers are non-emergent, resulting from a failure to progress in labor, the mother's desire for pain relief, or maternal exhaustion. Less common but more urgent indications include preeclampsia, maternal hemorrhage, retained placenta, malpresentation, sustained fetal distress, and respiratory problems in the newborn.

Third-party payers

Direct-entry midwives are reimbursed by private insurance plans and contract with managed care organizations in many states. A number of states mandate reimbursement under "every category of provider" or "any willing provider" laws. At least ten states reimburse direct-entry midwives for services provided to women on Medicaid. This is a high-priority issue for the National Association of Certified Professional Midwives (NACPM), which is committed to achieving national recognition for the CPM, including mandatory inclusion in Medicaid programs. The Patient Protection and Affordable Care Act 2010 stipulates Medicaid reimbursement for licensed care providers working in birth centers, which includes licensed midwives.

Integration Activities

NACPM is represented in numerous national organizations and initiatives focused on healthcare reform and women's health issues. These include the Integrated Healthcare Policy Consortium, Health Care for America Now, and the National Quality Forum. MANA is a member of the Coalition for Improving Maternity Services, a broad-based coalition of over 50 organizations, representing over 90,000 members. The coalition's mission is to promote a wellness model of maternity care that will improve birth outcomes and substantially reduce costs. MANA is also a partner in The Safe Motherhood Initiatives-USA, a partnership of organizations whose goal is to reduce maternal mortality in the United States. NACPM and MANA joined with four other groups to form the Midwives and Mothers in Action Campaign, a partnership whose goal is to gain federal recognition of certified professional midwives so that women and families will have increased access to quality, affordable maternity care in the settings of their choice.

In 2012, NACPM and the Association of Midwifery Educators co-hosted a symposium focused on the opportunities for integration and expansion of midwifery in a new maternity care system. A series of reports from the symposium were issued throughout 2012 and are available on the CPM Symposium website.

In 2011, a diverse group of stakeholders, including midwives and physicians, researchers and policy-makers, was invited to examine home birth within the context of the maternity care system. Using Future Search methodology to identify common ground and build consensus, the goal was to establish what the whole system can do to support those who choose homebirth, and provide the care, safety net, consultation, collaboration and referral necessary to make homebirth the safest and most positive experience for all involved: moms, babies, families, communities, healthcare workers, hospital personnel, administrators, payers, and so on.[12]

The following statements reflect the areas of consensus that were achieved by the individuals who participated in the Home Birth Consensus Summit:

STATEMENT 1

We uphold the autonomy of all childbearing women. All childbearing women, in all maternity care settings, should receive respectful, woman-centered care. This care should include opportunities for a shared decision-making process to help each woman make the choices that are right for her. Shared decision making includes mutual sharing of information about benefits and harms of the range of care options, respect for the woman's autonomy to make decisions in accordance with her values and preferences, and freedom from coercion or punishment for her choices.

STATEMENT 2

We believe that collaboration within an integrated maternity care system is essential for optimal mother-baby outcomes. All women and families planning a home or birth center birth have a right to respectful, safe, and seamless consultation, referral, transport and transfer of care when necessary. When ongoing interprofessional dialogue and cooperation occur, everyone benefits.

STATEMENT 3

We are committed to an equitable maternity care system without disparities in access, delivery of care, or outcomes. This system provides culturally appropriate and affordable care in all settings, in a manner that is acceptable to all communities. We are committed to an equitable educational system without disparities in access to affordable, culturally appropriate, and acceptable maternity care provider education for all communities.

STATEMENT 4

It is our goal that all health professionals who provide maternity care in home and birth center settings have a license that is based on national certification that includes defined competencies and standards for education and practice. We believe that guidelines should:

- allow for independent practice
- facilitate communication between providers and across care settings
- encourage professional responsibility and accountability; and,
- include mechanisms for risk assessment.

STATEMENT 5

We believe that increased participation by consumers in multi-stakeholder initiatives is essential to improving maternity care, including the development of high quality home birth services within an integrated maternity care system.

STATEMENT 6

Effective communication and collaboration across all disciplines caring for mothers and babies are essential for optimal outcomes across all settings. To achieve this, we believe that all health professional students and practitioners who are involved in maternity and newborn care must learn about each other's disciplines, and about maternity and health care in all settings.

STATEMENT 7

We are committed to improving the current medical liability system, which fails to justly serve society, families, and healthcare providers and contributes to:

- inadequate resources to support birth injured children and mothers
- unsustainable healthcare and litigation costs paid by all
- a hostile healthcare work environment
- inadequate access to home birth and birth center birth within an integrated healthcare system; and
- restricted choices in pregnancy and birth.

STATEMENT 8

We envision a compulsory process for the collection of patient (individual) level data on key process and outcome measures in all birth settings. These data would be linked to other data systems, used to inform quality improvement, and would thus enhance the evidence basis for care.

STATEMENT 9

We recognize and affirm the value of physiologic birth for women, babies, families and society and the value of appropriate interventions based on the best available evidence to achieve optimal outcomes for mothers and babies.[13]

In 2006, the White Ribbon Alliance for Safe Motherhood convened a national working group to develop guidelines to ensure that the healthcare needs of pregnant women, new mothers, fragile newborns, and infants would be adequately met during and after a disaster. They recommended that CPMs should be engaged in local and regional planning efforts, that home-birth skills should be taught to all providers, and that information be provided on how to prepare for birth outside the hospital.

In 2001 the American Public Health Association adopted a resolution supporting "Increasing Access To Out-Of-Hospital Maternity Care Services Through State-Regulated and Nationally-Certified Direct-Entry Midwives." APHA supports licensing and certification for direct-entry midwives, increased funding for scholarship and loan repayment programs, and eliminating barriers to the reimbursement and equitable payment of direct-entry midwives.[14]

In 1999 the Pew Health Professions Commission and the UCSF Center for the Health Professions issued a joint report on "The Future of Midwifery" calling for expanding educational opportunities for nurse-midwifery and direct-entry midwifery, health policies that facilitate integration of midwifery services, and research to evaluate practices.[15]

There are currently two midwifery education programs located in state-funded colleges and another in development. Faculties from other direct-entry programs are often invited to lecture in medical and nursing programs.

Education

Schools and Programs

There are currently nine schools and programs accredited by the Midwifery Education Accreditation Council. These include three degree-granting institutions, one program located within a university, and five schools offering certificate programs. Four offer distance learning courses in combination with clinical preceptorships. Three of the freestanding schools and the university-based program all participate in US Department of Education Title IV financial aid programs.

There are nearly 600 students enrolled in accredited midwifery programs and institutions. In addition, there are at least that many more independent students who are completing community-based apprenticeships, following guidelines provided by NARM, whose competency will be assessed individually through NARM's Portfolio Evaluation Process.

Accredited programs (July 2012):

- Bastyr University, Department of Midwifery, Kenmore, WA
  http://www.bastyr.edu/academics/schools-departments/school-natural-health-arts-sciences/department-midwifery

Accredited institutions (July 2012):

- Birthingway College of Midwifery, Portland, OR
  http://www.birthingway.edu/
- Birthwise Midwifery School, Bridgton, ME
  http://www.birthwisemidwifery.edu/
- Florida School of Traditional Midwifery, Gainesville, FL
  http://www.midwiferyschool.org
- Maternidad La Luz, El Paso, TX
  http://www.maternidadlaluz.com/
- Midwives College of Utah, Salt Lake City, UT
  http://www.midwifery.edu/
- National College of Midwifery, Taos, NM
  http://www.midwiferycollege.org/
- National Midwifery Institute, Bristol, VT
  http://www.nationalmidwiferyinstitute.com/
- Nizhoni Institute of Midwifery, San Diego, CA
  http://www.midwiferyatnizhoni.com/

Curriculum content

The *MANA Core Competencies for Basic Midwifery Practice* address the following content (in addition to the guiding principles mentioned elsewhere in this chapter):

I. General Knowledge and Skills

The midwife's knowledge and skills include but are not limited to:

A. communication, counseling and education before pregnancy and during the childbearing year
B. human anatomy and physiology, especially as relevant to childbearing
C. human sexuality
D. various therapeutic healthcare modalities for treating body, mind and spirit
E. community health care, wellness and social service resources
F. nutritional needs of the mother and baby during the childbearing year
G. diversity awareness and competency as it relates to childbearing

The midwife maintains professional standards of practice including but not limited to:

A. principles of informed consent and refusal and shared decision making
B. critical evaluation of evidence-based research findings and application to best practices
C. documentation of care throughout the childbearing cycle
D. ethical considerations relevant to reproductive health;
E. cultural sensitivity and competency

F. use of common medical terms
G. implementation of individualized plans for woman-centered midwifery care that support the relationship between the mother, the baby and their larger support community
H. judicious use of technology
I. self-assessment and acknowledgement of personal and professional limitations

II. Care during Pregnancy

The midwife provides care, support and information to women throughout pregnancy and determines the need for consultation, referral or transfer of care as appropriate. The midwife has knowledge and skills to provide care that include but are not limited to:

A. identification, evaluation and support of mother and baby well-being throughout the process of pregnancy
B. education and counseling during the childbearing cycle
C. identification of pre-existing conditions and preventive or supportive measures to enhance client well-being during pregnancy
D. nutritional requirements of pregnant women and methods of nutritional assessment and counseling
E. emotional, psychosocial and sexual variations that may occur during pregnancy
F. environmental and occupational hazards for pregnant women
G. methods of diagnosing pregnancy
H. the growth and development of the unborn baby
I. genetic factors that may indicate the need for counseling, testing or referral

J. indications for and risks and benefits of biotechnical screening methods and diagnostic tests used during pregnancy

K. anatomy, physiology and evaluation of the soft and bony structures of the pelvis

L. palpation skills for evaluation of the baby and the uterus

M. the causes, assessment and treatment of the common discomforts of pregnancy

N. identification, implications and appropriate treatment of various infections, disease conditions and other problems that may affect pregnancy

O. management and care of the Rh-negative woman

P. counseling to the woman and her family to plan for a safe, appropriate place for birth

III. Care during Labor, Birth and Immediately Thereafter
The midwife provides care, support and information to women throughout labor, birth and the hours immediately thereafter. The midwife determines the need for consultation, referral or transfer of care as appropriate. The midwife has knowledge and skills to provide care that include but are not limited to:

A. the processes of labor and birth

B. parameters and methods, including relevant health history, for evaluating the well-being of mother and baby during labor, birth and immediately thereafter

C. assessment of the birthing environment to assure that it is clean, safe and supportive and that appropriate equipment and supplies are on hand

D. maternal emotional responses and their impact during labor, birth and immediately thereafter

E. comfort and support measures during labor, birth and immediately thereafter

F. fetal and maternal anatomy and their interrelationship as relevant to assessing the baby's position and the progress of labor

G. techniques to assist and support the spontaneous vaginal birth of the baby and placenta

H. fluid and nutritional requirements during labor, birth and immediately thereafter

I. maternal rest and sleep as appropriate during the process of labor, birth and immediately thereafter

J. treatment for variations that can occur during the course of labor, birth and immediately thereafter, including prevention and treatment of maternal hemorrhage

K. emergency measures and transport for critical problems arising during labor, birth or immediately thereafter

L. appropriate support for the newborn's natural physiologic transition during the first minutes and hours following birth, including practices to enhance mother–baby attachment and family bonding

M. current biotechnical interventions and technologies that may be commonly used in a medical setting

N. care and repair of the perineum and surrounding tissues

O. third-stage management, including assessment of the placenta, membranes and umbilical cord

P. breastfeeding and lactation

Q. identification of pre-existing conditions and implementation of preventive or supportive measures to enhance client well-being during labor, birth, the immediate postpartum and breastfeeding

IV. Postpartum Care

The midwife provides care, support and information to women throughout the postpartum period and determines the need for consultation, referral or transfer of care as appropriate. The midwife has knowledge and skills to provide care that include but are not limited to:

A. anatomy and physiology of the mother
B. lactation support and appropriate breast care including treatments for problems with nursing
C. support of maternal well-being and mother-baby attachment
D. treatment for maternal discomforts
E. emotional, psychosocial, mental and sexual variations
F. maternal nutritional needs during the postpartum period and lactation
G. current treatments for problems such as postpartum depression and mental illness
H. grief counseling and support when necessary
I. family-planning methods, as the individual woman desires

V. Newborn Care

The midwife provides care to the newborn during the postpartum period, as well as support and information to parents regarding newborn care and informed decision making, and determines the need for consultation, referral or transfer of care as appropriate. The midwife's assessment, care and shared information include but are not limited to:

A. anatomy, physiology and support of the newborn's adjustment during the first days and weeks of life

B. newborn wellness, including relevant historical data and gestational age
C. nutritional needs of the newborn
D. benefits of breastfeeding and lactation support
E. laws and regulations regarding prophylactic biotechnical treatments and screening tests commonly used during the neonatal period
F. neonatal problems and abnormalities, including referral as appropriate
G. newborn growth, development, behavior, nutrition, feeding and care
H. immunizations, circumcision and safety needs of the newborn

VI. Women's Health Care and Family Planning

The midwife provides care, support and information to women regarding their reproductive health and determines the need for consultation or referral by using a foundation of knowledge and skills that include but are not limited to:

A. reproductive health care across the lifespan
B. evaluation of the woman's well-being, including relevant health history
C. anatomy and physiology of the female reproductive system and breasts
D. family planning and methods of contraception
E. decision making regarding timing of pregnancies and resources for counseling and referral
F. preconception and interconceptual care
G. well-woman gynecology as authorized by jurisdictional regulations

VII. Professional, Legal and Other Aspects of Midwifery Care
The midwife assumes responsibility for practicing in accordance with the principles and competencies outlined in this document. The midwife uses a foundation of theoretical knowledge, clinical assessment, critical-thinking skills and shared decision making that are based on:

A. MANA's Essential Documents concerning the art and practice of midwifery
B. the purpose and goals of MANA and local (state or provincial) midwifery associations
C. principles and practice of data collection as relevant to midwifery practice
D. ongoing education
E. critical review of evidence-based research findings in midwifery practice and application as appropriate
F. jurisdictional laws and regulations governing the practice of midwifery
G. basic knowledge of community maternal and child healthcare delivery systems
H. skills in entrepreneurship and midwifery business management

Accreditation

The Midwifery Education Accreditation Council (MEAC) is recognized by the US Secretary of Education as the accrediting agency for direct-entry midwifery education programs. All MEAC-accredited institutions and programs offer comprehensive programs that prepare graduates to become Certified Professional Midwives (CPMs). Most of the accredited programs are three years in length.

The standards for accreditation address curriculum requirements, faculty qualifications, facilities and students services, fiscal responsibility and administrative capacity. MEAC accreditation requires that the curriculum address core competencies and the essential knowledge and skills identified by NARM. The clinical component must be at least one year in duration and must include specific clinical experience requirements. Programs must provide evidence that graduates are capable of passing the NARM written examination and becoming CPMs and/or meeting the requirements for state licensure.

Regulation and Certification

Regulatory status

Direct-entry midwives are licensed or otherwise legally recognized practitioners in Alaska, Arkansas, Arizona, California, Colorado, Delaware, Florida, Idaho, Louisiana, Maine, Minnesota, Montana, New Hampshire, New Jersey, New Mexico, New York, Oregon, Rhode Island, South Carolina, Tennessee, Texas, Utah, Vermont, Virginia, Washington, and Wisconsin. All of these states (except New York) use the NARM written examination in their state regulatory process and/or recognize the CPM credential. A few states also require that applicants complete an accredited education program. Direct-entry midwives practice independently in most of these jurisdictions. Some state laws stipulate conditions requiring consultation or referral and others do not. Many provide a formulary of prescriptive drugs and devices that midwives may obtain and administer. Current information, including access to individual state laws, is available at the MANA website. [16]

In 2012, NARM adopted a position statement in support of state licensure for CPMs.[17] At that time, there were active efforts in at least 11 states to secure licensure legislation.

Examinations and certifications

The North American Registry of Midwives (NARM) is the certifying agency for CPMs. Certification is a credential that validates the knowledge, skills, and abilities vital to responsible midwifery practice, and that reflects and preserves the essential nature of midwifery care. The CPM credential is unique among maternity care providers in the US as it requires training and experience in out-of-hospital birth. The NARM certification process recognizes multiple routes of entry into midwifery and includes verification of knowledge and skills and the successful completion of both a written examination and skills assessment. NARM conducts periodic surveys of CPMs and completes a national job analysis to assure that the examination is based on real-life job requirements. Candidates for certification must be graduates of an accredited program or must complete a Portfolio Evaluation Process administered by NARM. Certification is renewed every three years and all CPMs must obtain continuing education and participate in peer review for recertification.

The CPM credential is accredited by the National Commission for Certifying Agencies (NCCA) which is the accrediting body of the National Organization for Competency Assurance (NOCA).

NCCA encourages their accredited certification programs to have an education evaluation process so candidates who have been educated outside of established pathways may have their qualifications evaluated for credentialing. The NARM Portfolio Evaluation Process meets this recommendation.

Certified professional midwives must demonstrate that they have met the minimum education, skills, and experience requirements set forth by the North American Registry of Midwives. Competence is assessed against a comprehensive and detailed list of knowledge, skills and abilities. Students must also participate in a minimum of 55 births (at least 5 births in homes and 2 births in hospital), including 25 births where the student functions in the role of primary midwife under supervision, providing full continuity of care for at least 5 women. Students must also participate in specific minimum numbers of prenatal, postpartum and newborn care experiences.

Research

There is a growing body of evidence that points to the many benefits of midwifery care. In October 2008, the Milbank Memorial Fund released *Evidence-Based Maternity Care: What It Is and What It Can Achieve.* The report presented best evidence that, if widely implemented, would have a positive impact on many mothers and babies and would improve value for payers. The report also identified barriers to providing evidence-based maternity care, and made policy recommendations to address the barriers. Care provided by certified professional midwives, documented in a landmark study of CPMs published in 2005, was highlighted in the Milbank Report:

> "The low CPM rates of intervention are benchmarks for what the majority of childbearing women and babies who are in good health might achieve."

An international review of the literature published by the Cochrane Collaboration in October 2008 found that "Midwife-led care was associated with several benefits for mothers and babies, and had no identified adverse effects. The main benefits were a reduced risk of losing a baby before 24 weeks. Also during labor, there was a reduced use of regional analgesia, with fewer episiotomies or instrumental births. Midwife-led care also increased the woman's chance of being cared for in labor by a midwife she had got to know. It also increased the chance of a spontaneous vaginal birth and initiation of breastfeeding. In addition, midwife-led care led to more women feeling they were in control during labor. There was no difference in risk of a mother losing her baby after 24 weeks." The review concluded that all women should be offered midwife-led models of care.

Another comprehensive review of the scientific evidence underlying the Coalition for Improving Maternity Services (CIMS), "Ten Steps of Mother-Friendly Care," was published in 2007. The authors found that equally good or better outcomes can be achieved in low-risk women having planned home births or giving birth in freestanding birth centers.

An economic cost-benefit analysis of direct-entry midwifery licensing and practice was undertaken in 2007 at the request of the Washington State Department of Health. The researchers determined that planned out-of-hospital births attended by professional midwives had similar rates of intrapartum and neonatal mortality to those of low-risk hospital births. They also found that medical intervention rates for planned out-of-hospital births were lower than planned low-risk hospital births. Using conservative cost estimates, they estimated the cost savings to the healthcare system (public and private insurance) was $2.7 million per biennium and recovery from Medicaid fee-for-service alone was more than $473,000 per biennium. These cost savings were achieved by licensed midwives attending just 2% of all births in the state.

Midwives Alliance of North America data collection has been ongoing since 1993 and includes a complete set of prospective data contributed by CPMs during the year 2000, when NARM required their participation. MANA established a Division of Research in 2004; it is charged with administering an online data-entry system, establishing research priorities, promoting publication, and providing education on research concepts and methods. A new web-based data collection system was launched in November 2004 and allows midwives to enter and review their own practice data as well as compiling data from all practices. Mechanisms are in place to provide access to the database for researchers to use this unique system to study various aspects of midwifery care. Experts from the Centers for Disease Control have shown an interest in this database as it may now be the only repository of normal birth data in the country.

Challenges and Opportunities

Key challenges 2012—2015

- coordinate an effective response to the increasing medicalization of pregnancy and birth, including the rapidly rising cesarean-section rate
- eliminate disparities in perinatal outcomes by addressing the role that midwives can play and promoting the benefits of midwifery care
- preserve and promote the principles, values, and practices of midwifery within the context of continuing professsion-alization of the home-birth and midwifery movements
- secure legislation in states where midwifery is not recognized or is currently illegal
- support faculty development, particularly training for preceptors and researchers
- increase the number of students of color and midwife instructors of color as a key strategy in increasing the number of midwives and the impact of midwifery care
- build capacity and develop funding to support new and ongoing professional organization activities

Key opportunities 2012—2015

- public awareness and demand for midwifery services is increasing; support continued public education and outreach, including specific campaigns launched by MANA and the American College of Nurse-Midwives (ACNM)

- healthcare reform includes a number of federal initiatives to improve maternity care; leverage these to increase recognition, expand job opportunities and third party reimbursement for direct-entry midwives
- home birth services and birth centers are gaining recognition as a legitimate, high-quality and cost-effective element of maternity care; build on the consensus statements from the 2011 Home Birth Consensus Summit to improve integration and support for women and their care providers in all settings
- workforce needs are changing and a shortage of maternity care providers is predicted; develop funding sources and other support to increase the number of direct-entry midwifery education programs and expand opportunities for training
- health professional education, like higher education more generally, is changing rapidly to meet the needs of a new era; encourage innovation to improve access to cutting-edge learning opportunities
- midwifery research opportunities are growing; support evaluation of midwifery practices to expand the knowledge base and to inform women's choices in maternity care and decisions by health policy-makers, particularly the plan to complete another round of mandatory prospective data collection from CPMs in 2015
- global standards for midwifery education and regulation have been set by the International Confederation of Midwives; utilize these tools to create common understanding and build unity among all midwifery organizations in the US

Resources

Organizations and websites

The National Association of Certified Professional Midwives (NACPM) was incorporated in 2000 to "empower and care for women as they make choices about their pregnancies and births, through the development and growth of the profession of midwifery." The NACPM strives to ensure that midwifery as practiced by CPMs will take its rightful place in the delivery of maternity care to women and families in the United States. Participation in healthcare reform and federal recognition of the CPM are current priorities for the NACPM.

The Midwives Alliance of North America (MANA) was founded in 1982 to form an identifiable and cohesive organization representing the profession of midwifery. MANA supported the creation of national standards for the certification of midwives. The certification process is administered by the North American Registry of Midwives and results in the credential Certified Professional Midwife (CPM). However, MANA does not require that members be certified and has an ongoing commitment to be an inclusive organization accessible to all midwives, regardless of their credential or legal status. MANA oversees numerous projects, including legislative advocacy, international relations, and public relations. The MANA Division of Research maintains a web-based system for prospective data collection that is a resource to midwives and other researchers interested in the processes and outcomes of midwifery care.

The Association of Midwifery Educators was launched in 2006 to strengthen schools and support midwifery educators. They publish a newsletter which has addressed topics ranging from faculty qualifications to student admissions practices. They also host a website and provide numerous opportunities for networking.

- American Association of Birth Centers
  www.birthcenters.org/
- American College of Nurse-Midwives
  www.acnm.org/
- Association of Midwifery Educators
  www.associationofmidwiferyeducators.org/
- Childbirth Connection
  www.childbirthconnection.org/
- Citizens for Midwifery
  www.cfmidwifery.org/
- Coalition for Improving Maternity Services
  www.motherfriendly.org/
- CPM Symposium
  www.cpmsymposium.com/
- Foundation for the Advancement of Midwifery
  www.foundationformidwifery.org/
- International Confederation of Midwives
  www.internationalmidwives.org/
- Midwifery Education Accreditation Council
  www.meacschools.org/
- Midwives Alliance of North America
  www.mana.org/
- National Association of Certified Professional Midwives
  www.nacpm.org/
- North American Registry of Midwives
  www.narm.org/

Bibliography
Selected books about midwifery and maternity care
Bridgman Perkins, B. *The Medical Delivery Business: Health Reform, Childbirth and the Economic Order.* Piscataway, NJ: Rutgers University Press; 2004.

Davis-Floyd R, Johnson CB, eds. *Mainstreaming Midwives: The Politics of Change.* New York, NY: Routledge; 2006.

Gaskin IM. *Ina May's Guide to Natural Childbirth.* New York, NY: Bantam; 2003.

Rooks JP. *Midwifery and Childbirth in America.* Philadelphia, PA: Temple University Press; 1999.

Rothman BK. *Laboring On: Birth in Transition in the United States.* New York, NY: Routledge; 2007.

Wagner M. *Born in the USA: How a Broken Maternity System Must Be Fixed to Put Women and Children First.* Berkeley, CA: University of California Press; 2006.

Reference texts
Cunningham FG, Gant NF, Leveno KJ, Gilstrap LC, Hauth JC, Wenstrom, KD. *William's Obstetrics.* 21st ed. New York, NY: McGraw-Hill; 2001.

Davis E. *Heart and Hands: A Midwife's Guide to Pregnancy and Birth.* 4th ed. Berkeley, CA: Celestial Arts; 2004.

Enkin M, Keirse M, Neilson J, et al. *A Guide to Effective Care in Pregnancy and Birth.* New York, NY: Oxford University Press; 2000.

Frye A. *Holistic Midwifery: A Comprehensive Textbook for Midwives and Home Birth Practice. Volume 1 – Care During Pregnancy.* Portland, OR: Labrys Press; 1995.

Frye A. *Holistic Midwifery: A Comprehensive Textbook for Midwives and Home Birth Practice. Volume 2 – Care During Labor and Birth.* Portland, OR: Labrys Press; 2004.

Frye A. *Understanding Diagnostic Tests in the Childbearing Year.* Portland, OR: Labrys Press; 1997.

Gaskin IM. *Spiritual Midwifery.* 4th ed. Summertown, TN: The Book Publishing Company; 2002.

Myles M. *Textbook for Midwives.* 14th ed. Philadelphia, PA: Elsevier; 2000.

Oxhorn H. *Human Labor and Birth.* 5th ed. New York, NY: McGraw-Hill; 1986.

Page LA. *The New Midwifery.* Philadelphia, PA: Churchill Livingstone; 2000.

Renfrew M, Fisher C, Arms S. *Bestfeeding: Getting Breastfeeding Right.* 2nd ed. Berkeley, CA: Celestial Arts; 2000.

Simkin P. *Labor Progress Handbook.* Cambridge, MA: Blackwell Scientific; 2000.

Sinclair C. *A Midwife's Handbook.* St. Louis, MO: Saunders; 2004.

Thureen P. *Assessment and Care of the Well Newborn.* St. Louis, MO: Saunders; 1998.

Varney H, Kriebs JM, Gegor CL. *Varney's Midwifery.* 4th ed. Sudbury, MA: Jones and Bartlett; 2004.

Weaver P, Evans S. *Practical Skills Guide for Midwifery.* 3rd ed. Wasilla, AK: Morningstar Publishing; 2001.

Wickham S. *Midwifery Best Practice.* Philadelphia, PA: Elsevier; 2003.

Reports and recommendations

Lamaze International. The Coalition for Improving Maternity Services: Evidence Basis for the Ten Steps of Mother-Friendly Care. *Supp J Perinat Educ.*2007;16(1).

The Mother-Friendly Childbirth Initiative. Coalition for Improving Maternity Services Web site. http://www.motherfriendly.org/mfci.php. 1996. Accessed November 17, 2009.

Wells S, Nelson C, Kotch JB, et al. Increasing Access to Out-of-Hospital Maternity Care Services through State-Regulated, Nationally-Certified Direct-Entry Midwives. Midwives Alliance of North America Website.

http://mana.org/APHAformatted.pdf. 2001. Accessed
November 17, 2009

Midwifery education, professional associations, state regulation

Bourgeault I, Fynes M. Integrating lay and nurse-midwifery into
the US and Canadian healthcare systems. *Soc Sci Med.*
1997;44(7):1051-1063.

Suarez SH. Midwifery is not the practice of medicine. *Yale J Law
Fem.* 1993;5(2):315-364.

Scope of practice, standards of practice, and quality assurance

Spindel PG, Suarez SH. Informed consent and home birth. *J
Nurse Midwifery.* 1995;40(6):541-552.

Suarez SH. Midwifery is not the practice of medicine. *Yale J Law
Fem.* 1993;5(2):315-364.

Consultation and referral

Davis-Floyd R. Home-birth emergencies in the US and Mexico:
the trouble with transport. *Soc Sci Med.* 2003;56:1911-1931.

Stapleton SR. Team-building: making collaborative practice
work. *J Nurse Midwifery.* 1998;43(1):12-18.

Citations

1. International Confederation of Midwives. International
   definition of the midwife.
   http://www.internationalmidwives.org/who-we-are/policy-
   and-practice/icm-international-definition-of-the-midwife/
2. Core Competencies for Basic Midwifery Practice. August 4,
   2011. Midwives Alliance North America.
   http://mana.org/pdfs/MANACoreCompetenciesColor.pdf
3. Hatem M, Sandall J, Devane D, Soltani H, Gates S. Midwife-
   led versus other models of care for childbearing women.
   Cochrane Database Syst Rev. 2008;(4).

4. Sakala C, Corry M. *Evidence-Based Maternity Care: What It Is and What It Can Achieve*. New York, NY: Milbank Memorial Fund; 2008.
5. Kennedy H. A model of exemplary midwifery practice: results of a Delphi study. *J Midwifery Womens Health*. 2000;45(4):4-19.
6. Rooks, JP. The midwifery model of care. *J Nurse Midwifery*. 1999;44(4):370-374.
7. Janssen P, Lee SK, Ryan EM, et al. Outcomes of planned home births versus planned hospital births after regulation of midwifery in British Columbia. *CMAJ*. 2002;166(3):315-23.
8. Johnson KC, Daviss BA. Outcomes of planned home births with certified professional midwives: large prospective study in North America. *BMJ*. 2005; 330(7505):1416.
9. Health Management Associates. Midwifery Licensure and Discipline Program in Washington State: Economic Costs and Benefits. http://washingtonmidwives.org/documents/Midwifery_Cost _Study_10-31-07.pdf , Accessed July 4, 2012.
10. Sullivan DA, Weitz R. *Labor Pains: Modern Midwives and Home*. New Haven/London: Yale University Press;1988:68-80
11. North American Registry of Midwives. What is a CPM? www.narm.org/
12. Home Birth Consensus Summit, 2011. What Was the Process Engaging in dialogue: The Future Search model. http://www.homebirthsummit.org/history/what-was-the-process , Accessed July 4, 2012
13. Home Birth Consensus Summit, 2011. Common Ground. http://www.homebirthsummit.org/outcomes/common-ground-statements , Accessed July 4, 2012
14. American Public Health Association. Increasing Access to Out-of-Hospital Maternity Care Services through State-Regulated and Nationally-Certified Direct-Entry Midwives. January 1, 2001. Policy Number: 20013.

http://www.apha.org/advocacy/policy/policysearch/default.h
tm?id=242

15. Dower CM, Miller JE, O'Neil EH and the Taskforce on
    Midwifery. Charting a Course for the 21st Century: The
    Future of Midwifery. San Francisco, CA: Pew Health
    Professions Commission and the UCSF Center for the Health
    Professions. April 1999.
    http://www.futurehealth.ucsf.edu/Content/29/1999-
    04_Charting_a_Course_%20for_the_21st_Century_The_Futur
    e_of_Midwifery.pdf

16. Midwives Alliance of North America. Direct-Entry
    Midwifery State-by-State Legal Status. May 11, 2011.
    http://www.mana.org/statechart.html

17. NARM North American Registry of Midwives. State
    Licensure of Certified Professional Midwives. April, 2012.
    http://narm.org/wp-content/uploads/2012/05/State-
    Licensure-of-CPMs2012.pdf

Naturopathic Medicine

First Edition (2009) Authors: Paul Mittman, ND, EdD, Patricia Wolfe, ND, Michael Traub, ND, DHANP, FABNO

Second Edition (2013) Editors: Christa Louise, MS, PhD; Paul Mittman, ND, EdD; Michael Traub, ND, DHANP, FABNO; Marcia Prenguber, ND, FABNO; Elizabeth Pimentel, ND

Partner Organization: Association of Accredited Naturopathic Medical Colleges (AANMC)

About the Authors/Editors: Mittman is President of Southwest College of Naturopathic Medicine and past President of the Association of Accredited Naturopathic Medical Colleges. Wolfe is Director Emeritus of the Boucher Institute of Naturopathic Medicine and a past Association of Accredited Naturopathic Medical Colleges member. Traub is a private practitioner and a past President of the American Association of Naturopathic Physicians. Louise is the Executive Director of the North American Board of Naturopathic Examiners and is an ACCAHC Board member. Prenguber is the Director of Integrative Care for Indiana University Health Goshen Center for Cancer Care, is the ACCAHC Board member nominated by the Council on Naturopathic Medical Education, and is Co-chair of the ACCAHC Clinical Working Group. Pimentel is the Dean of the University of Bridgeport College of Naturopathic Medicine, is the ACCAHC Board member nominated by the Association of Accredited Naturopathic Medical Colleges, and is an ACCAHC Education Working Group member.

Philosophy, Mission, and Goals

Although naturopathic medicine as an organized system of practice was first conceived at the beginning of the 20th century, it has its roots in traditions that date back to the time of Hippocrates. Naturopathic medicine is a holistic approach to health care that recognizes and respects the individuality of the patient. The American Association of Naturopathic Physicians (AANP) defines naturopathic medicine as "a distinct system of primary health care: an art, science, philosophy, and practice of diagnosis, treatment, and prevention of illness."

Naturopathic medicine has its foundation in the biomedical sciences, and is a comprehensive system of health care. Naturopathic physicians act to stimulate the inherent self-healing abilities of the individual through lifestyle change and the application of non-suppressive therapeutic methods and modalities, including clinical nutrition, botanical medicine, homeopathy, physical medicine, and health psychology. Naturopathic physicians are also schooled in and able to apply conventional clinical practices including emergency medicine, minor surgery, pharmacology, and natural childbirth.[1] Naturopathic physicians respect and apply the principles of traditional world practices such as Traditional Chinese Medicine and Ayurvedic medicine, which provide an added basis for understanding the individual patient from a constitutional perspective.

The techniques of naturopathic medicine include traditional and modern, empirical and scientific methods.[2] Naturopathic medicine is distinguished by the principles upon which its practice is based. The application of these principles in practice is continually reexamined in the light of scientific advances.

The philosophy of naturopathic medicine is embodied by six principles:

First, Do No Harm (primum no nocere)
Naturopathic physicians strive to minimize harmful side effects, using the least invasive means necessary to diagnose and treat the patient.

Use the Healing Power of Nature (vis medicatrix naturae)
Naturopathic physicians recognize and seek to strengthen the inherent ability of the body to heal itself. The naturopathic physician's role is to work with the patient to identify and remove obstacles to healing, and to facilitate and enhance the self-healing process.

Identify and Treat the Cause
Naturopathic physicians strive to identify and address the underlying causes of disease.

Treat the Whole Person
Naturopathic physicians recognize that the health of the patient must be addressed at all levels: physical, mental, emotional, spiritual, social, and environmental.

Educate the Patient (docere)
A primary role of the naturopathic physician is to educate the patient and empower the individual to take responsibility for her/his own health. Naturopathic physicians recognize and foster the therapeutic power of the doctor/patient partnership.

Focus on Prevention
Naturopathic physicians believe that preventing disease is preferable to waiting until the disease must be treated. This focus includes assessing risk factors and susceptibility to disease, encouraging lifestyle choices that prevent

diseases from manifesting, and intervening in disease processes to slow their progression.

Naturopathic medicine is aimed at strengthening the body's natural self-regulatory (self-healing) ability. In the naturopathic model, therapies are applied to reestablish physiological and psychological homeodynamics. Symptoms are seen as the body's way of alerting the patient to pay attention to some aspect of her/his health. Naturopathic physicians believe that there are detrimental health consequences to merely suppressing symptoms without addressing the cause. For example, the naturopathic model recognizes fever as the body's attempt to heal itself. Naturopathic treatment might be aimed at stimulating the fever (while monitoring the process to prevent an elevation that would be dangerous), with the knowledge that when the temperature has reached a certain level, the body's self-regulatory system will be activated and the trajectory of the temperature rise will be reversed. The naturopathic model posits that even specific diseases will benefit from non-specific therapies, because these non-specific therapies will enhance the body's ability to self-regulate. The use of homeopathy in naturopathic medicine is exemplary of this assertion that when the body is given the information it needs, it will self-correct.

Several key features of the naturopathic model distinguish it from the allopathic model:

1. The distinction between disease (as a physiological process) and illness (as the psychological experience of the condition) is de-emphasized in the naturopathic model. The term *imbalance* is used to refer to both the biological aspects that initiate or exacerbate the problem and to the patient's experience. Both the biological aspects and the patient's experience must be addressed.

2. The naturopathic model is holistic. It recognizes the interconnectedness of physiological systems (e.g., the central nervous system and the immunological system) and of systems at different levels of the organism (e.g., the physical level and the psychological level). Naturopathic medicine features a multifactorial approach to healing (e.g., a treatment plan that includes specific therapeutics, diet, exercise, and stress management).

3. Naturopathic medicine works within the framework of self-regulation, i.e., it works with the biological systems that have been organized to maintain health. This is apparent in the belief that the body has the power to heal itself and that the role of the physician is to remove obstacles to that healing.

4. Naturopathic medicine works within as well as between levels of the organism. Healing at the deepest levels cannot occur until healing has taken place at the more superficial levels. Symptoms that are suppressed without addressing the cause will later manifest in other ways, potentially at deeper levels which will lead to more serious disease states. Therefore, healing must be addressed at all levels.

5. Because naturopathic philosophy acknowledges multifactorial etiology to dysfunction, naturopathic treatment is highly individualized, taking into account the unique circumstances of the patient's physiology and life. Treatment is specifically targeted toward the individual patient's particular areas of dysregulation.

6. The patient-physician relationship is seen as a partnership, with the patient collaborating in treatment. This relationship is recognized as being a vital aspect of the healing process.

7. Quality is as important as quantity. This applies to the foods the patient puts into her/his body (nutrient-rich or nutrient-empty), the kind of air s/he breathes (clean or polluted), the environment in which s/he lives (toxin-free or toxic), and the

kind of relationships in her/his life (supportive or destructive), etc.

8. Naturopathic medicine recognizes and seeks to enhance three kinds of effects. *Specific* effects result from the application of a therapy that is known to have a particular effect on a targeted area (e.g., ingestion of a botanical substance that is known to reduce gastric acidity). *General* effects result from individualized treatment prescriptions, and serve to increase the body's resistance to external disruptions (e.g., the use of Astragalus to stimulate the immune system). *Non-specific* effects are those effects that are a result of the mind/body connection.

Characteristics and Data

In 2000, the University of California, San Francisco Center for the Health Professions reported that there were approximately 2,000 licensed naturopathic physicians practicing in North America[3]. Since then, the profession has more than doubled. In 2012, the American Association of Naturopathic Physicians (AANP) estimated that there were approximately 3,400 naturopathic physicians licensed/registered in the United States and its territories, and the Canadian Association of Naturopathic Doctors estimated that there were approximately 1,900 naturopathic doctors licensed/registered in the Canadian provinces and territories. Of the respondents in the most recent (2011) Naturopathic Physicians Licensing Examinations (NPLEX) Practice Analysis (PA) survey, 77% were female, the mean age was 41 (median 39), and 81% were Caucasian (with 6% Asian, 5% multi-ethnic, 1% Latino, 1% Black/African American). Of the survey respondents, approximately 20% were practicing in unregulated states/provinces.[4]

According to the *Chronicle of Higher Education's* 2009 Occupational Brief,[5] naturopathic physicians work 30 to 50 hours a week. Of the respondents in the 2011 NPLEX PA survey, 63% said they work at least 30 hours per week. Most naturopathic physicians provide health care through office-based practice. Practitioners may make house calls or see patients in the evenings or on weekends. In addition to seeing patients, some naturopathic physicians serve as adjunct faculty at one of the naturopathic medical colleges or other health professions programs (e.g., nursing). Many lecture in other venues.

While entry-level naturopathic physicians typically earn $35,000 a year with the expectation of annual increases, some new graduates are challenged to earn adequate income if they are not strong entrepreneurs. The NPLEX PA survey found that of the naturopathic doctors who were working less than full time, 50% cited lack of patients as the reason.[4] Experienced naturopathic physicians may earn from $45,000 to $100,000 a year. The NPLEX PA Survey found that approximately 10% of the respondents reported *gross* earnings of more than $200,000 per year,[4] but how this translates to *net* income is dependent on many factors, including the number of hours devoted to practice, the location of the practice, overhead expenses, and insurance coverage. Earning potential is also affected by professional certification, the types of services offered, the emphasis of treatment, and the population where the practice is centered.

Clinical Care

Approach to patient care
Using the six principles outlined above, naturopathic physicians seek to understand the factors that underlie and contribute to the patient's condition, to remove obstacles to healing, to strengthen the patient's inherent healing ability, and to teach the patient about dietary and lifestyle choices that support wellness and optimal health.[6] In striving to attain an in-depth understanding of the patient's health, naturopathic physicians consider genetic

predispositions in combination with superimposed factors such as nutritional status, work and emotional stressors, environmental allergens and toxins, and bio-mechanical issues. Essential to a comprehensive evaluation is the extended interview, which ranges from 60 to 90 minutes for a new patient. A standard review of systems is supplemented with patient-generated reports of daily activities (including dietary habits, physical activity, and psychological well-being). Naturopathic physicians perform physical examinations and use conventional as well as innovative laboratory and diagnostic imaging procedures when appropriate. With many treatment options at their disposal, naturopathic physicians follow a distinct clinical rationale to individualize patient care within the framework of naturopathic principles. Follow-up appointments typically range from 30 to 45 minutes, and may occur more frequently in the beginning in order to support dietary and lifestyle changes.

In 1996 the AANP recommended that further work on practice principles move from the profession to the academic community. Clinical faculty and practitioners built on the core foundation throughout the 1990s. Three principles were identified that underlie the *clinical practice* of naturopathic medicine: [7,8]

1. Disease is characterized as a process rather than a pathologic entity;
2. focus is on the determinants of health rather than on pathology; and
3. a therapeutic hierarchy guides the course of treatment.

As taught in naturopathic medical schools, the therapeutic hierarchy is a guideline for applying various modalities and approaches according to the unique needs of the individual patient and the natural order of the healing process. The therapeutic hierarchy proposes the following order from most gentle to most invasive:

1. establish the conditions for health by removing obstacles to healing
2. stimulate the self-healing mechanisms
3. support weakened or damaged systems or organs
4. address structural integrity
5. address pathology using specific natural substances, modalities, or interventions
6. address pathology using specific pharmacologic or synthetic substances
7. employ surgical correction and other invasive therapies that may have significant side effects.

For example, for a child who has recurrent otitis media, conventional treatment would consist of repeated courses of antibiotics and, in some cases, tympanostomy. A naturopathic physician would take a different approach:

1. Use the child's medical history, physical examination, and laboratory tests to look for and remove or address obstacles to health. In this child's case, these obstacles may include allergens (e.g., dust mites, dander, etc.), environmental irritants (e.g., second hand cigarette-smoke), food sensitivities (dairy, wheat, and eggs are most often implicated), poor nutrition (e.g., a diet high in processed grains and sugars), mechanical misalignments (e.g., of the cervical spine or cranial bones), and emotional factors (e.g., family stress). Children routinely improve significantly after these factors are addressed.
2. Stimulate the healing power of nature with therapies such as homeopathy and hydrotherapy.

3. Strengthen the affected systems by providing general immune support with nutritional supplementation (e.g., vitamins C and A) or botanical medicines (e.g., Echinacea angustifolia).
4. Address structural factors (e.g., using soft tissue manual therapy such as lymphatic drainage).
5. Treat the pathology with specific natural therapies (e.g., topical garlic, *Verbascum Thapsus* oil, *Hypericum perforatum* oil).
6. Prescribe a course of antibiotics if the infection is particularly recalcitrant or severe, or if there are other compromising factors.
7. Refer the child to a pediatrician or otolaryngologist if the infection still does not subside.

The therapeutic hierarchy creates a guideline for treatment that is both consistent with the principles of naturopathic medicine and addresses the patient's dynamic needs.[9]

Scope of practice

Naturopathic physicians are trained as primary care providers (PCPs). Some choose to specialize in populations, modalities, or other focused areas of clinical practice. Every jurisdiction that licenses/regulates the practice of naturopathic medicine gives naturopathic physicians the authority to diagnose as well as treat. This ability to diagnose distinguishes naturopathic physicians from practitioners who use natural therapies (such as botanical medicines and homeopathy) but who were not trained at CNME-approved naturopathic medical programs and who are not licensable in any of the regulated jurisdictions. Diagnosis includes eliciting a medical history, performing a physical examination, and (in many cases) ordering laboratory tests and/or diagnostic imaging studies.

In the states that have functioning regulatory Boards, naturopathic physicians are allowed to prescribe botanical medicines, nutritional supplements, and homeopathic remedies, to provide physical therapies, and to provide health and lifestyle counseling. In the past 3 years, several states have expanded their formularies and scopes of practice. Eight states allow naturopathic physicians to practice minor surgery, and eleven states have drug prescription rights that allow some type of parenteral (IM or IV) administration of supplements or drugs. In some states, prescriptive authority is limited to first-line antibiotics and bio-identical hormones, while in other states naturopathic physicians have full prescriptive authority for all drugs except chemotherapeutics and anti-psychotics. Prescribing rights are important for continuity of care, as a naturopathic physician cannot take a patient off a drug unless s/he also has the right to prescribe that drug. Several states allow for expanded scope of practice in specific areas under the naturopathic license. For example, three states allow naturopathic doctors to practice acupuncture, and 9 states allow them to practice natural childbirth (primarily in out-of-hospital settings).

Referral practices

Naturopathic physicians are trained and may serve as primary care providers. They collaborate with other health professionals and refer patients for optimum management of the patient's health care, especially for those patients who have life-threatening or challenging chronic conditions.[10]

Third-party payers

Insurance reimbursement for naturopathic physicians in the United States varies from jurisdiction to jurisdiction, although in the regulated states many private and governmental programs cover naturopathic care. The NPLEX PA survey found that only

19% of patient visits in North America were covered by insurance, although it must be kept in mind that not all naturopathic physicians accept third-party insurance. Some states provide student loan forgiveness to naturopathic physicians for service to urban and rural underserved communities and include naturopathic physicians as primary care providers in medical home legislation. In Vermont, naturopathic care is covered under Medicaid. In Oregon, naturopathic physicians are included in some coordinated care organizations. In Canada, reimbursement varies from province to province, but minimal coverage is available in some provincial healthcare programs. Unfortunately, restrictions on fees and numbers of covered visits drive most naturopathic doctors out of provincial healthcare programs. Private insurance companies in Canada are beginning to offer more reasonable coverage, but most of the population does not have private insurance.

Integration Activities

Naturopathic physicians practice in a wide variety of clinical settings and collaborate with diverse practitioners from both conventional and alternative health professions. The NPLEX PA survey found that only 32% of the respondents were practicing in an office by themselves; 22% practiced in clinics with other members of their profession, 22% practiced in clinics with other non-ND healthcare professionals, and 24% practiced in clinics where there were other naturopathic doctors as well as practitioners from other professions. Examples of integration with more conventional medical practitioners include:

- Integrative oncology: A number of oncology centers around the country integrate naturopathic physicians into their hospital and outpatient centers, providing patients with a wide range of treatment options and expertise in combating their illness.

- Integrative rheumatology: Many patients suffering from autoimmune diseases such as rheumatoid arthritis and systemic lupus erythematosus benefit by combining the skills and knowledge of a conventional rheumatologist (MD) and a naturopathic physician.
- Two Institute of Medicine projects in 2009-2011 included naturopathic physicians either on committees or contracted with them for papers.
- Public, rural, and community health: public and community health clinics that are in proximity to naturopathic colleges often integrate services from naturopathic physicians.
- For international communities in need (e.g., Nicaragua, Mexico, Africa, India), healthcare services are being provided through naturopathic college-based student preceptorships and internships offered by organizations such as Naturopathic Doctors International and Naturopathic Medicine for Global Health.

Education

Schools and Programs
Approximately 2250 students are currently enrolled at one of seven CNME-accredited naturopathic medical schools:

- Bastyr University in Seattle, Washington, founded in 1978
    - o Including Bastyr University branch campus in San Diego, California, opened in 2012
- Boucher Institute of Naturopathic Medicine in Vancouver, British Columbia, founded in 1998
- Canadian College of Naturopathic Medicine in Toronto, Ontario, founded in 1978

- National College of Natural Medicine in Portland, Oregon, founded in 1956
- National University of Health Sciences in Lombard, Illinois, Naturopathic Program started in 2004
- Southwest College of Naturopathic Medicine in Tempe, Arizona, founded in 1992
- University of Bridgeport College of Naturopathic Medicine in Bridgeport, Connecticut, Program started in 1997

Student Population

For students enrolled in the Fall 2011 entering class, the mean cumulative GPA was 3.30, and the mean science GPA was 3.46. The mean age was 31, with a bi-modal distribution of students in their early 20s continuing directly after their undergraduate programs, and older students pursuing a second or third career. Minority students comprised approximately 22% of the 2011 entering class. The typical naturopathic student population is approximately 80% female.

Faculty

Faculty in the biomedical sciences predominantly have PhDs. Clinical faculty at naturopathic medical schools must have a terminal degree in their respective fields. The clinical sciences are taught by naturopathic doctors and other licensed practitioners such as medical doctors, osteopathic doctors, chiropractors, psychologists, and practitioners of acupuncture and Oriental medicine.

Curriculum content

Naturopathic physicians are trained through an educational process that is comparable to that of allopathic medical doctors (MDs) and doctors of osteopathy (DOs). Acceptance into a naturopathic medical school requires a bachelor's degree with pre-medical sciences, including biology, general and organic chemistry, and physics, as well as psychology and humanities.

The naturopathic college admission process requires official undergraduate transcripts and essays that provide insight into the candidate's understanding of naturopathic medicine, as well as her/his motivation for applying to naturopathic medical school. Individuals who meet the application criteria are interviewed by faculty, staff, and students to assess academic ability, interpersonal skills, and professional demeanor.

With the biomedical sciences as a foundation, naturopathic medical education integrates Western diagnostic decision-making skills with natural as well as conventional therapies. Students spend more than 4000 hours during the four-year post baccalaureate program learning the art and science of naturopathic medicine. Naturopathic principles, philosophy, and theory guide the curriculum and provide a conceptual framework for students to develop a profound understanding of humans in health and disease. The curriculum is dynamic, adapting to meet the changing healthcare needs of the population and to address the many obstacles to healing that patients experience. Naturopathic medical education pays particular attention to the growing epidemic of chronic diseases that affect every age group and sector of society.

At naturopathic medical school, students take two years of graduate level studies in the biomedical sciences. Although naturopathic medical students receive training in the same biomedical sciences in which allopathic medical students are trained, there are differences in the emphasis in the courses and in the way that knowledge is applied. For example, slightly less emphasis is placed on anatomy, requiring an average of 350 hours in a naturopathic curriculum, compared to 380 hours in an allopathic/osteopathic curriculum. However, in naturopathic programs, greater emphasis is placed on physiology. Naturopathic students receive approximately twice as many hours of physiology as allopathic students (250 hours compared to 125 hours)[11]. This difference may reflect the greater emphasis in naturopathic medicine on working with the dynamic processes of the person, rather than on the static structure of the body.

Coursework during the first year focuses on developing students' understanding of the human body in health and disease. Recent trends show programs moving toward integrated systems-based biomedical training, with clinical education introduced in the first year or two. Anatomy, biochemistry, microbiology, physiology, embryology, histology, neuroanatomy, and genetics accompany introductory classes in naturopathic philosophy, nutrition, mind-body medicine, homeopathy, and botanical medicine. The second year introduces students to Western diagnostic knowledge, with courses in clinical diagnosis, pathology, lab diagnosis, and diagnostic imaging. Lab sections teach students critical skills in history-taking and physical examination. Intermediate courses in naturopathic therapeutics continue to deepen students' understanding of clinical nutrition, manipulative therapy and physical medicine, homeopathy, and botanical medicine. In years three and four, students engage in supervised patient care on diverse patient populations. Didactic education continues to build expertise in naturopathic therapeutics, and adds in-depth coursework in pediatrics, gynecology, gastroenterology orthopedics, cardiovascular health, disorders of the eyes, ears, nose, and throat, oncology, nephrology, and dermatology. Naturopathic principles guide the curriculum design and course content.

Clinical training takes place in a variety of clinical settings. Clinical faculty provide training and supervision of care in multidisciplinary medical centers. Students see patients who have a wide range of clinical conditions, from acute illnesses (e.g., respiratory infections, influenza, gastrointestinal infections, musculoskeletal injuries, and minor lacerations) to chronic, sometimes life-threatening diseases (e.g., asthma, diabetes, colitis, heart disease, hypertension, hyperlipidemia, metabolic syndrome, arthritis, cancer, kidney disease, etc.). Experienced licensed clinicians, including naturopathic doctors, medical doctors, osteopaths, chiropractors, psychologists, and practitioners of acupuncture and Oriental medicine supervise students and oversee high quality patient care in the school clinics. In addition to the colleges' outpatient clinics and medical centers, students provide free care to thousands of medically underserved women, children, and men at homeless shelters, HIV/AIDS clinics, drug and alcohol rehabilitation programs, elementary schools in impoverished immigrant communities, shelters for battered women and children, and community health centers. These programs benefit the patients, who receive high quality naturopathic, conventional, and preventive care, as well as the students, who are introduced to a diverse population of patients who are often under-insured and suffering disproportionately from both chronic and acute conditions.

Accreditation
Founded in 1978, the Council on Naturopathic Medical Education (CNME) was incorporated in the District of Columbia to accredit naturopathic medical programs in the United States and Canada. A member of the Association of Specialized and Professional Accreditors, the CNME is the only naturopathic accrediting body recognized by the US Department of Education.

The CNME Board comprises institutional members from accredited naturopathic medical schools, profession members who are licensed naturopathic physicians and public members who possess educational experience and are unaffiliated with any naturopathic program or the profession. In addition to establishing educational criteria for accreditation, the CNME requires programs to participate in the self-study process. Site visits are conducted for candidacy, for initial accreditation, and to renew accreditation.[12]

In addition to programmatic accreditation by the CNME, all naturopathic medical schools in the United States have institutional accreditation from their respective regional accrediting agencies and in Canada from their provincial accrediting bodies. Accreditation holds institutions to high academic, financial, and human resource standards. Accreditation also ensures that schools respect academic freedom in teaching and include faculty, staff, and students in decision-making processes. Equally important, accredited institutions join a community of colleges and universities who share the ideas, innovations, and experiences that drive continuous quality improvement.

As noted above, seven naturopathic programs are currently accredited by the CNME.

Regulation and Certification

Regulatory status

The Federation of Naturopathic Medicine Regulatory Authorities was established in 2011 to facilitate communication between regulatory bodies, and to serve as a single body with which other entities in the profession can communicate. In the United States, naturopathic medicine is regulated in 16 states (Alaska, Arizona, California, Connecticut, Idaho, Hawaii, Kansas, Maine, Minnesota, Montana, New Hampshire, North Dakota, Oregon, Utah, Vermont, and Washington) and the District of Columbia. Two territories, Puerto Rico and the Virgin Islands, are also regulated. Licensure efforts are underway in at least 12 other states. In Canada, naturopathic medicine is regulated in six provinces (Alberta, British Columbia, Manitoba, Nova Scotia, Ontario, and Saskatchewan). A high priority for both the American Association of Naturopathic Physicians and the Canadian Association of Naturopathic Doctors is to support legislative efforts to increase the number of states and provinces that regulate the profession.

· To be eligible to be licensed/registered as a naturopathic physician in a regulated jurisdiction, a candidate must graduate from an accredited naturopathic medical school, and must pass national board examinations and other jurisdiction-specific examinations.

Board examinations

The North American Board of Naturopathic Examiners (NABNE) is responsible for ensuring that candidates for licensure/registration have the requisite knowledge and skills to be safe practitioners. NABNE administers the Naturopathic Physicians Licensing Examinations (NPLEX), which are the board examinations required by all jurisdictions that regulate naturopathic medicine.

NPLEX is administered in two parts. Both parts are competency-based. Students take the Part I Biomedical Science Examination after they complete their biomedical science training. Part II Clinical Sciences Examinations may be taken after the candidate has passed the Part I Examination and has graduated from a CNME-accredited naturopathic medical school.

NPLEX Part I—Biomedical Science Exam Areas

Anatomy
Biochemistry & Genetics
Microbiology & Immunology
Pathology
Physiology

NPLEX Part II— Clinical Science Examinations
Part II Core Clinical Science Exam Areas:
Physical & Clinical Diagnosis
Laboratory Diagnosis & Diagnostic Imaging
Botanical Medicine
Clinical Nutrition
Physical Medicine
Homeopathy
Health Psychology & Counseling
Research
Emergency Medicine & Medical Procedures
Pharmacology
Part II—Elective Examinations
Minor Surgery
Acupuncture

The clinical examinations are based on a practice analysis of the profession, the most recent of which was completed in 2011. Both the Part I and Part II Examinations use an integrated case-based format. For the Part I Biomedical Science Examination, a brief case is presented and the examinee is asked four or five questions regarding anatomy, physiology, biochemistry, pathology, microbiology and/or immunology that are relevant to the patient's diagnosis. For the Part II Core Clinical Science Examination, a more complete case is presented (including results of lab testing and diagnostic imaging), and the examinee is asked to diagnose the condition, interpret lab and imaging findings, and prescribe appropriate treatments within the context of naturopathic principles.

Research

Clinical trials and some laboratory studies are conducted at naturopathic medical schools in the United States and Canada.

As the profession matures and the capacity for research expands (reflected in institutional commitment and development of a skilled research workforce), the number and quality of research studies improve. Naturopathic doctors often earn an MPH (Masters in Public Health) or PhD degree as part of their development as researchers, and then work at naturopathic institutions or at conventional academic health centers. There are currently naturopathic doctor researchers at the University of Washington, Yale University, the University of Michigan, and Oregon Health and Science University. Others work with agencies such as the Group Health Research Institute, RAND Corporation, and the Samueli Institute.

From 2002-2004, the National Institutes of Health (NIH) National Center for Complementary and Alternative Medicine (NCCAM) funded the development of the Naturopathic Medical Research Agenda (NMRA). The NMRA brought together naturopathic physicians and conventional research scientists to identify and prioritize a list of research questions and areas.[13] Research departments from all naturopathic medical schools participated in a series of meetings and produced a consensus document on the future of naturopathic research. The highest priorities focused on three areas:

- Naturopathic treatment of type 2 diabetes
- Naturopathic care for the preservation and promotion of optimal health in geriatric populations
- Developing methodologies to understand the healing process

Between 1999 and 2012, NCCAM awarded approximately $25 million in research grants to institutions with programs in naturopathic medicine. Studies have been published on naturopathic approaches to specific conditions (e.g., diabetes and cardiovascular disease), the effectiveness of individual botanical medicines (e.g., *Urtica urens* for hay fever, *Echinacea angustifolia* for URIs, *Cimecifuga racemosa* for menopausal symptoms); homeopathic medicines (e.g., coca for altitude sickness, regional pollens for hay fever); nutritional supplementation (e.g., SAMe for osteoarthritis) and naturopathic detoxification for environmental illnesses. Other published studies have examined the nature of patient experience with naturopathic treatment.

In 2010, the Naturopathic Physicians Research Institute (NPRI) was founded. A key focus in this organization has been "researching the way we practice" using research methods which recognize that most disease has multiple causes and is best addressed using an array of interventions. Virtually all naturopathic practice reflects this perspective; consequently, single agent trials do not adequately measure the value of naturopathic medicine or support enhancement of its practice. The NPRI website lists some of these emerging whole practice trials. Naturopathic researchers were among the leaders who played a significant role in a successful ACCAHC-led effort to educate the NCCAM to elevate the importance of examining the impact of whole disciplines in the agency's 2011-2015 strategic plan.

Challenges and Opportunities

Key challenges 2012-2015

- The greatest challenge facing the naturopathic profession today is the insufficient number of states and provinces that regulate naturopathic medicine. Currently less than a third of the states regulate the profession. With more than 95% of licensed naturopathic physicians practicing in only 16 regulated states, the need for what naturopathic medicine has to offer cannot be met nationally. Furthermore, healthcare consumers in the 34 unregulated states do not have adequate knowledge of naturopathic medicine. This has implications not only for patient demand, but for student recruitment as well. The profession must continue to push for regulation in more states and provinces.

- Consumers are confused regarding the differences between licensed naturopathic doctors and those practitioners (in unregulated states) who call themselves naturopaths or NDs but who do not have professional degrees or the credentials to be safe and effective practitioners. The profession must focus on educating consumers about the value of trained, licensed naturopathic physicians.
- Scope of practice in some jurisdictions is limited.
- Although the number of naturopathic medical schools has increased significantly, there are still fewer than 6,000 naturopathic physicians in practice today. The small number of naturopathic practitioners clearly limits the opportunity to play a more significant role in the growing popular demand for the integrative medical practices in which naturopathic physicians are trained.

Key opportunities 2012-2015

The capacity to train naturopathic physicians has increased significantly (the number of naturopathic medical schools has more than doubled in the past 20 years), and graduating class sizes are increasing. There is a larger and better qualified applicant pool available to the schools. Competencies and outcomes have been clearly defined for accredited schools and are regularly updated as the profession develops. All of these factors position the profession to take advantage of the opportunities that are presenting themselves, including:

- There is a growing need for primary care providers, particularly in light of the increased number of Americans who have access to health care due to the 2010 Affordable Care Act. The naturopathic profession's greatest strength lies in its ability to help address this need. Furthermore, the number of conventional medical students pursuing primary care/family care has been decreasing (down to 24% in Canada), leaving a void which NDs can help fill.

- There is an emerging epidemic of chronic diseases traceable to diet, environment, and lifestyle (e.g., diabetes, hypertension, heart disease, gastrointestinal disorders, environmental illnesses, allergies, etc.). Recognition by even the conventional healthcare profession of these etiologic factors is making it clear that patients must accept responsibility for their own health; it is vital that patients have physicians with whom they can partner on this journey. Naturopathic physicians are trained to work with their patients, spending time to educate them on lifestyle factors. Consequently, the services naturopathic physicians have to offer are in greater demand than ever before.

- There are diverse practice opportunities for naturopathic physicians that were unheard of twenty years ago. Naturopathic physicians increasingly work in clinical settings with medical and osteopathic doctors, chiropractors, acupuncturists, massage therapists, and other CAM practitioners. Naturopathic physicians can be found in tribal clinics on Native American lands, in public health clinics, in cancer treatment centers, in integrative medical centers, in mental health clinics, in substance abuse programs, in corporate wellness programs, and at resorts and spas.

- Collaboration is increasing among schools and other professional ND organizations for the growth and advancement of the profession and for greater accessibility for patients.

- State and national healthcare reform are providing opportunities for greater integration of naturopathic physicians into the mainstream healthcare system.

Resources

Organizations and websites

National Associations:
- American Association of Naturopathic Physicians www.naturopathic.org
- Canadian Association of Naturopathic Doctors www.cand.ca (also www.naturopathicassoc.ca)

Education and Research:
- Association of Accredited Naturopathic Medical Colleges www.aanmc.org
- Council on Naturopathic Medical Education www.cnme.org
- Bastyr University www.bastyr.edu
- Boucher Institute of Naturopathic Medicine www.binm.org
- Canadian College of Naturopathic Medicine www.ccnm.edu
- National College of Natural Medicine www.ncnm.edu
- National University of the Health Sciences www.nuhs.edu
- Southwest College of Naturopathic Medicine www.scnm.edu
- University of Bridgeport College of Naturopathic Medicine www.bridgeport.edu/academics/graduate/naturo
- Naturopathic Post-Graduate Association

www.np-ga.org
- Naturopathic Medical Student Association
  www.naturopathicstudent.org
- Naturopathic Physicians Research Institute
  www.nprinstitute.org

Regulation:
- Federation of Naturopathic Medicine Regulatory Authorities
  www.fnmra.org
- North American Board of Naturopathic Examiners
  www.nabne.org

Specialty Associations:
- American Association of Naturopathic Midwives
  www.naturopathicmidwives.com
- Homeopathic Academy of Naturopathic Physicians
  www.hanp.net
- Oncology Association of Naturopathic Physicians
  www.oncanp.org
- Pediatric Association of Naturopathic Physicians
  www.pedanp.org
- Institute of Naturopathic Generative Medicine
  www.generativemedicine.org/INGM

Other Organizations:
- Natural Doctors International
  www.ndimed.org
- Naturopathic Medicine for Global Health
  www.natmedglobalhealth.org

Bibliography

General Naturopathic Medicine:

Alschuler LN, Gazella KA. *Alternative Medicine Magazine's Definitive Guide to Cancer: An Integrative Approach to Prevention, Treatment, and Healing.* Berkeley, CA: Celestial Arts; 2007.

Bove M. *An Encyclopedia of Natural Healing for Children and Infants.* New York, NY: McGraw-Hill; 2001.

Canadian College of Naturopathic Medicine Press. Fifteen textbooks based on naturopathic curriculum, written by naturopathic physicians, faculty, and lecturers. http://www.ccnmpress.com/. Accessed November 17, 2009.

Hudson T. *Women's Encyclopedia of Natural Medicine: Alternative Therapies and Integrative Medicine.* Lincolnwood, IL: Keats; 1999.

Pizzorno JE, Murray MT, Joiner-Bey H. *The Clinician's Handbook of Natural Medicine.* 2nd ed. Philadelphia, PA: Churchill Livingston; 2002.

Pizzorno JE, Murray MT. *The Encyclopedia of Natural Medicine.* 2nd ed, revised. Roseville, CA: Prima Publishing; 1997.

Pizzorno JE, Murray MT. *The Textbook of Natural Medicine.* 4th ed. Philadelphia, PA: Elsevier; 2013.

Pizzorno JE. *Total Wellness: Improve Your Health by Understanding the Body's Healing Systems.* Roseville, CA: Prima Publishing; 1997.

Standish L, Calabrese C, Snider P. *The Future and Foundations of Naturopathic Medical Research Science: Naturopathic Medical Research Agenda.* Kenmore, WA: Bastyr University Press; 2005.

Yarnell E. *Naturopathic Gastroenterology.* East Wenatchee, WA: Healing Mountain Publishing Inc; 2011.

Naturopathic Philosophy:

Boyle W, Saine A. *Lectures in Naturopathic Hydrotherapy.* East Palestine, OH: Buckeye Naturopathic Press; 1988.

Kirchfeld F, Boyle W. *Nature Doctors: Pioneers in Naturopathic Medicine.* Portland, OR: Medicina Biologica; 1994.

Kneipp S. *My Water-Cure.* New ed. Champaign, IL: Standard Publications Inc; 2007.

Lindlahr H. *Nature Cure.* Charleston, SC: BiblioBazaar LLC; 2007.

Lindlahr H. *Philosophy of Natural Therapeutics.* Champaign, IL: Standard Publications Inc; 2007.

Lust B. *Collected Works of Dr. Benedict Lust.* East Wenatchee, WA: Healing Mountain Publishing Inc; 2006.

Smith F. *An Introduction to Principles and Practice of Naturopathic Medicine.* Toronto, Ontario, Canada: CCNM Press; 2008.

Zeff JL. The process of healing: a unifying theory of naturopathic medicine. *J Naturopathic Med.* 1997;7(1):122-5.

Zeff J , Snider P, Myers S, DeGrandpre Z. A hierarchy of healing: the therapeutic order. The unifying theory of naturopathic medicine. In: Pizzorno JE, Murray MT. *Textbook of Natural Medicine.* 4th ed. Philadelphia, PA: Elsevier;2013, 18-33.

Other:

Chaitow L, Blake E, Orrock P, Wallden M, Snider P, Zeff J. *Naturopathic Physical Medicine: Theory and Practice For Manual Therapists and Naturopaths.* Philadelphia, PA: Churchill Livingston Elsevier; 2008.

Dooley TR. *Homeopathy Beyond Flat Earth Medicine.* 2nd ed. San Diego, CA: Timing Publications; 2002.

Marz RB. *Medical Nutrition from Marz.* Portland, OR: Quiet Lion Press; 1997.

Mitchell W. *Plant Medicine in Practice: Using the Teachings of John Bastyr.* Philadelphia, PA: Churchill Livingston Publishing; 2003.

Murray M, Pizzorno J, Pizzorno L. *The Condensed Encyclopedia of Healing Foods.* New York, NY: Atria Books; 2006.

Neustadt J. *A Revolution in Health through Nutritional Biochemistry.* Bloomington, IN: iUniverse, Inc; 2007.

Stargrove M, Treasure J, McKee DL. *Herb, Nutrient, and Drug Interactions: Clinical Implications and Therapeutic Strategies.* St. Louis, MO: Mosby; 2007.

Tilgner S. *Herbal Medicine from the Heart of the Earth.* Pleasant Hill, OR: Wise Acres; 1999.

Journals and Periodicals:
International Journal of Naturopathic Medicine. http://www.intjnm.org. Peer reviewed online journal.

Natural Medicine Journal. www.naturalmedicinejournal.com/. Peer reviewed online journal.

Naturopathic Doctor News and Review. http://www.ndnr.com. Professional news and information resource for naturopathic physicians in North America.

Citations
1. Hough H, Dower C, O'Nell E. *Profile of a Profession: Naturopathic Practice.* San Francisco, CA: University of California at San Francisco Center for the Health Professions; 2001.
2. American Association of Naturopathic Physicians. Select Committee on the Definition of Naturopathic Medicine, Snider P, Zeff J, co-chairs. Definition of Naturopathic Medicine. Position Paper. Rippling River, OR; 1989.
3. Albert D, Martinez D. The supply of naturopathic physicians in the United States and Canada continues to increase. *Comp Health Prac Rev.* 2006; 11(2):120-122.
4. Naturopathic Physicians Licensing Examinations (NPLEX) 2011 Practice Analysis of the Naturopathic Profession. Portland, OR. Mountain Measurement Inc. http://www.mountainmeasurement.com/contact.php
5. Chronicle Guidance Publications. Occupational Brief 624: Naturopathic Physicians.

http://www.chronicleguidance.com/store.asp?pid=7330 Accessed November 17, 2009.

6. Dunne N, Benda W, Kim L, et al. Naturopathic medicine: what can patients expect? *J Fam Pract.* 2005;54(12).

7. Zeff J. The process of healing: a unifying theory of naturopathic medicine. *J Naturopathic Med.* 1977;7(1):122-125.

8. Zeff J , Snider P, Myers S, DeGrandpre Z. A hierarchy of healing: the therapeutic order. The unifying theory of naturopathic medicine. In: Pizzorno JE, Murray MT. *Textbook of Natural Medicine.* 4th ed. Philadelphia, PA: Elsevier; 2013, 18-33.

9. Findings and Recommendations Regarding the Prescribing and Furnishing Authority of a Naturopathic Doctor. California. Dept. of Consumer Affairs, Bureau of Naturopathic Medicine; January 2007. http://www.naturopathic.ca.gov/formspubs/formulary_report.pdf

10. Health Professions Advisors Guide. Champaign, IL: National Association of Advisors for the Health Professions Inc; 2007.

11. Jensen, C B. (1997). Common paths in medical education: the training of allopaths, osteopaths, and naturopaths. *Alternative and Complementary Therapies.*1997;3(4):276-280.

12. US Department of Education Office of Postsecondary Education. http://www.ed.gov/admins/finaid/accred/index.html Accessed November 17, 2009.

13. Standish L, Calabrese C, Snider P. The naturopathic medical research agenda: the future and foundation of naturopathic medical science. *J Altern Complement Med.* 2006;12(3):341-345.

Section II

Related Integrative Practice Fields

Ayurvedic Medicine

Holistic Medicine

Holistic Nursing

Homeopathy

Integrated Medicine

Yoga Therapy

Ayurvedic Medicine

First Edition (2009) Author and Second Edition (2013) Editor:
Felicia Marie Tomasko, RN

Partner Organization: National Ayurvedic Medical Association
(NAMA)

About the Author/Editor: Tomasko is an Ayurvedic professional
and a member of the Board of Directors of the National
Ayurvedic Medical Association, a member of the Board of
Directors of the California Association of Ayurvedic Medicine, a
member of the Board of Directors of ACCAHC and the Editor-
in-Chief of *LA YOGA Ayurveda and Health* magazine.

Philosophy, Mission, and Goals

While Ayurveda as a system is in its relative infancy in the US as
far as education, regulation, recognition, scope, and numbers of
people practicing, it has a long historical tradition in its region of
origin, the Indian subcontinent. A translation of the Sanskrit
word Ayurveda is "science of life." The written source texts of
this science are variously dated but are generally agreed to date
to at least the beginning of the Common Era. Many scholars
suggest the written tradition is older, or at least a compilation of
a more ancient oral tradition. The three primary texts are the
Charaka Samhita, the *Sushruta Samhita*, and *Vagbhata's Ashtanga
Hridaya Samhita*. Information included therein is considered to be
the genesis of classical Ayurveda. Other texts have added to the
body of knowledge over time.

Ayurveda focuses on an individual's relationship
with his/her own body, mind, and spirit, and with the natural
world. These relationships begin with the five elements (earth,

water, fire, air, and ether/space) and how they combine to create the three *doshas*: *vata* (air and ether/space), *pitta* (fire and water), and *kapha* (water and earth). Vata is light, dry, expansive, and changeable. Pitta is hot, intense, transformational, and sharp. Kapha is stable, heavy, solid, cold, and oily. Many physiological processes fall under these categories; movement is governed by vata, digestion and transformation by pitta, and insulation and lubrication by kapha.

According to Ayurveda, people have individual constitutional make-ups (*prakruti*), with unique proportionality of the doshas. When in balance, good health is experienced. When out of balance, disease is more likely to take hold. Importance is placed on disease prevention and maintaining balance with the ever-changing forces of nature. Balance is not static, but is constantly shifting; individuals must adjust their practices, routines, and lifestyles to maintain balance. Ayurveda is not exclusive of modern medicine, and combines well philosophically and practically because of its inclusivity.

Characteristics and Data

The National Ayurvedic Medical Association (NAMA) is currently collecting information on people practicing Ayurveda in the US. This data has not yet been publically released.

Scope of Practice
Defining scope of practice related to the modification of the traditional textual tradition to comply with modern integrative medical practice is a current project of NAMA. Updated communication related to the ongoing discussions can be found at: http://www.ayurvedanama.org/?page=StandardsCommittee

Integration Activities

Many Ayurvedic professionals are also licensed healthcare providers in fields such as medicine, midwifery, chiropractic, nursing, massage therapy, and physical therapy. Many Ayurvedic professionals are also yoga teachers and/or yoga therapists. Concepts of Ayurveda are taught in yoga teacher training programs and are part of the core curriculum in the standards of the International Association of Yoga Therapists. Diverse collaborations are occurring in research settings, clinics, and hospitals.

Education

Ayurveda became initially visible in the US largely mainly through efforts of spiritual communities. In the 1970s, the Transcendental Meditation (TM) community brought doctors trained in India to the US to offer Ayurvedic consultations to meditators. Esteemed doctors, including Indian natives who came to the US to practice, Westerners who studied in India, and practitioners who trained in the West, have furthered the profession in the US.

People practicing Ayurveda have varying educational backgrounds, including BAMS (bachelor of Ayurvedic medical science) degrees, MASc (masters in Ayurvedic science) degrees or, more recently, MD Ayurved (medical doctor of Ayurveda) degrees. Training programs exist outside the Indian system of education including certificate courses and master's level and clinical doctorate programs. Each type enrolls and graduates students in the US. In addition to Western-style academic programs, all schools feature an internship component; some schools also have traditional apprenticeship models in addition to academic studies (above and beyond in-person internship required at all schools). Currently, schools are growing and expanding their internship programs. Representatives from 30 US-based schools have met to discuss the development of educational standards and scope of practice. As of 2012, 21 US-based schools have registered with NAMA, meet current requirements and participate in the discussions on standards.

Regulation and Certification

While Ayurveda is a licensed medical profession in its native India, and in countries including Nepal and Sri Lanka, Ayurveda is currently not licensed in the US. Yet self-regulation of the profession is increasing as organizations, such as NAMA, work on standards. Several state organizations (in California, Colorado, Florida, Minnesota and Washington) have missions and goals involving licensure.

Research

A growing number of studies confirm the efficacy of many traditional treatments, such as the use of turmeric (*Curcuma longa*), a spice valued in Ayurveda for its anti-inflammatory and blood sugar regulating effects and bitter gourd (*Momordica charantia*) for diabetes prevention and treatment.

The Transcendental Meditation community sponsors hundreds of scientific studies investigating the use of meditation. Research into yoga therapy, which overlaps with Ayurveda, is on the rise. The study of Ayurvedic therapies using a Western approach is still developing when compared to other systems, and faces challenges as therapies are generally tailored to the individual rather than standardized for a particular condition.

Challenges and Opportunities

Key challenges 2012-2015
Greater understanding and recognition of Ayurveda by consumers and the larger medical community are current challenges. Others include:

- identification of Ayurvedic practitioners at all levels currently active in the US
- unification of a dispersed and diverse community
- development of clear educational standards that take into account differing training methodologies and lineages within Ayurveda
- acceptance of drafted scope of practice guidelines for three different levels of practitioner with different amounts of training and ability to work with individuals and groups
- continued development of scope of practice guidelines that remain true to Ayurveda's long historical tradition and are relevant in modern integrated healthcare practice
- strengthening educational programs to train practitioners to work with other healthcare providers
- development and dissemination of research priorities and methodologies
- communication among the Ayurvedic communities worldwide to develop common goals in furthering the profession
- coordination within the herbal community to support sustainability in obtaining farmed or gathered herbs as well as supporting appropriately labeled and safe imported herbal supplies

Key opportunities 2012-2015

Practices supporting the achievement of balance and optimal health can contribute to making a positive difference in the modern healthcare system. Opportunities include:

- teaching tangible, low-cost, self-implemented techniques for health maintenance and possible prevention of chronic diseases
- increased collaboration with other healthcare providers and professions
- combining the philosophy and therapies of Ayurveda with other systems to address often recalcitrant chronic illnesses
- continued support of healthcare freedom legislation as well as pursuing licensure
- increased membership in NAMA as a means of networking and uniting the profession

Resources

Organizations and websites

The National Ayurvedic Medical Association is a 501(c)(6) organization and the largest Ayurvedic professional organization in the US. Membership is approximately 500. NAMA holds an annual conference. NAMA has a council of schools that focuses on standards and scope issues. While NAMA may be the most visible organization of Ayurvedic practitioners, it does not necessarily represent everyone within the US Ayurvedic community. Website: www.ayurveda-nama.org

Some states have active organizations, notably:

- California Association of Ayurvedic Medicine (CAAM) www.ayurveda-caam.org/

- Washington Ayurvedic Medical Association (WAMA) www.ayurveda-wama.org/
- Colorado Ayurvedic Medical Association (Colorama) http://www.coloradoayurveda.org/

Holistic Medicine

First Edition (2009) Authors: Hal Blatman, MD; Kjersten Gmeiner, MD; Donna Nowak, CH, CRT

Second Edition (2013) Editors: Molly Roberts, MD, MS; Steve Cadwell

Partner Organization: American Holistic Medical Association (AHMA)

About the Authors/Editors: Blatman is a past Board President, Gmeiner previously served as a Trustee, and Nowak is a past Executive Director for the American Holistic Medical Association. Roberts is the current Board President, and Cadwell currently serves as Executive Director of the American Holistic Medical Association.

Philosophy, Mission, and Goals

Holistic medicine is the art and science of healing that addresses care of the whole person—body, mind, and spirit. Since 1978, the American Holistic Medical Association (AHMA) has been helping transform conventional/allopathic medicine to a more holistic model. The practice of holistic medicine integrates conventional health care with complementary and alternative medicine (CAM) therapies to promote optimal health and wellness as well as to prevent and treat disease. The AHMA serves as an advocate for the use of holistic and integrative medicine by all licensed healthcare providers. We are the home for the collective voices of both those providing and those seeking more integrative healthcare.

1. **Optimal Health** is the primary goal of holistic medical practice. It is the conscious pursuit of the highest level of functioning and balance of the physical, environmental, mental, emotional, social and spiritual aspects of human experience, resulting in a dynamic state of being fully alive. This creates a condition of well-being regardless of the presence or absence of disease.

2. **The Healing Power of Love**. Holistic healthcare practitioners strive to meet the patient with grace, kindness, acceptance, and spirit without condition, as love is life's most powerful healer.

3. **Whole Person.** Holistic healthcare practitioners view people as the unity of body, mind, spirit and the systems in which they live.

4. **Prevention and Treatment**. Holistic healthcare practitioners promote health, prevent illness and help raise awareness of dis-ease in our lives rather than merely managing symptoms. A holistic approach relieves symptoms, modifies contributing factors, and enhances the patient's life system to optimize future well-being.

5. **Innate Healing Power.** All people have innate powers of healing in their bodies, minds and spirits. Holistic healthcare practitioners evoke and help patients utilize these powers to affect the healing process.

6. **Integration of Healing Systems.** Holistic healthcare practitioners embrace a lifetime of learning about all safe and effective options in diagnosis and treatment, which are selected to best meet the unique needs of the patient. The realm of choices may include lifestyle modification and complementary approaches as well as conventional drugs and surgery.

7. **Relationship-centered Care.** The ideal practitioner-patient relationship is a partnership which encourages patient autonomy, and values the needs and insights of both parties. The quality of this relationship is an essential contributor to the healing process.
8. **Individuality.** Holistic healthcare practitioners focus patient care on the unique needs and nature of the person who has an illness rather than the illness that has the person.
9. **Teaching by Example.** Holistic healthcare practitioners continually work toward the personal incorporation of the principles of holistic health, which then profoundly influence the quality of the healing relationship.
10. **Learning Opportunities.** All life experiences including birth, joy, suffering and the dying process are profound learning opportunities for both patients and healthcare practitioners.

Characteristics and Data

Holistic medicine in the US was conceived and practiced by medical doctors as early as the 1950s, building on the various complementary medical traditions of the US dating back to the mid-19th century. Currently, more than 1,600 physicians (MDs and DOs) have been certified by the American Board of Integrative Holistic Medicine (ABIHM).

Scope of Practice
Scope of practice is defined by conventional medical licensing statutes and agencies in all 50 states. Three states, Arizona, Connecticut, and Nevada, have homeopathic medical boards for medical doctors practicing homeopathy and other complementary and alternative therapies.

Integration Activities

- Participation in cross-discipline dialogues such as the Summit in Humanistic Medicine and the National Education Dialogue to Advance Integrated Health Care
- Full voting membership in the AHMA includes MDs, DOs, and all licensed healthcare providers
- Collaborative conferences (e.g. iMosaic in April 2011; Rapid Therapeutic Response in March 2012; Gateway to Health in April 2013)
- Development of new approaches for specialty complementary and integrative medicine education (including CME) as well as the development of local programming across the country
- Enhancement of AHMA's website to support those who provide services in the field of integrative health care
- AHMA Board representation from all categories of licensed healthcare providers who are members as well as other business specialties and the public
- AHMA is a part of a larger organization, Integrative Medicine Consortium (IMC), which is an umbrella group for eight organizations.

Education

Many holistic physicians receive their graduate training in another field and then become board-certified in holistic medicine. The American Board of Integrative Holistic Medicine (ABIHM) offers a board review course for physicians, using a defined, evidence-based curriculum. Conventional MD or DO training and licensing is the prerequisite for this board certification in holistic medicine. Many postgraduate fellowships are available in integrative medicine. A list of 21 fellowships is available here: http://www.abpsus.org/integrative-medicine-fellowships.

A new residency in integrative medicine—Integrative Medicine in Residency (IMR) program—is currently being piloted in eight family practice residency programs. It is currently in year three of its curriculum. IMR is a required, in-depth competency-based curriculum that is integrated into a typical three-year family practice residency program. It utilizes a web-based curriculum, program-specific experiential exercises, and group process-oriented activities.

Regulation and Certification

Board certification has been available via examination through the American Board of Integrative Holistic Medicine (ABIHM) since 1999. For more information, visit www.holisticboard.org. The ABIHM is currently developing the American Board of Integrative Medicine (ABOIM) with the University of Arizona and other academic institutions under the administration of the American Board of Physician Specialties (ABPS). The ABOIM certification will involve a fellowship prerequisite, which may be waived for physicians who have studied/practiced integrative medicine. Active ABIHM Diplomates will be partially eligible for a fellowship waiver. For more information visit: www.holisticboard.org and www.abpsus.org/integrative-medicine.

Research

Studies of clinical sites practicing integrative/holistic medicine are underway to examine the impact of specific modalities and therapies. AHMA serves as a resource to individuals and organizations doing research on systems of practice.

According to The Bravewell Collaborative, a philanthropic organization that works to improve health care, over the past two decades there has been documented growth in the number of clinical centers providing integrative medicine, the number of medical schools teaching integrative studies, the number of researchers studying integrative interventions, and the number of patients seeking integrative care.

Challenges and Opportunities

Key challenges 2012-2015

- Referring patients to AHMA-member licensed practitioners
- Advancing the principles of holistic medicine into the conventional medical model
- Maintaining economic stability of the organization and the field
- Continuing educational opportunities, which may include events, credits, and webinars
- Establishing legitimacy amid the current medical hierarchy
- Establishing clear procedure for proper licensing and training

Key opportunities 2012-2015

Through an internal restructuring and major educational campaign, AHMA expects to continue to increase membership over the next three years. Other opportunities include:

- Connecting doctors together through member gatherings and retreats
- Increasing the understanding and stability of the field of holistic medicine through the strengthening of interdisciplinary and inter-organizational relationships
- Transforming holistic medicine practitioners through conferences and workshops
- Bringing holistic medicine to communities across the country
- Increasing education of healthcare consumers

- Increasing the number of physicians who are adequately trained and Board certified through the ABOIM

Resources

Organizations

Part of a wider organization, Integrative Medicine Consortium (IMC), the American Holistic Medical Association (AHMA) was founded in 1978 as a 501(c)(3) nonprofit membership organization for physicians seeking to practice a broader form of medicine than what was (and is) currently taught in allopathic and osteopathic (MD and DO) medical schools. For more than 34 years, the AHMA has nurtured and educated physicians making this transition. The current membership includes approximately 800 doctors (including MDs, DOs, NDs, DCs, and others), medical students, residents, and licensed healthcare professionals (e.g., nurses, physician assistants, acupuncturists, nutritionists, massage therapists). AHMA is helping transform health care through collaboration, communication, and education.

www.holisticmedicine.org
27629 Chagrin Blvd., Suite 213, Woodmere, Ohio 44122,
Phone: (216) 292-6644, Fax: (216) 292-6688.

Holistic Nursing

First Edition (2009) Author and Second Edition (2013) Editor:
Carla Mariano, EdD, RN, AHN-BC, FAAIM

Partner Organization: American Holistic Nurses Association (AHNA)

About the Author/Editor: Carla Mariano developed and is former coordinator of the Adult Holistic Nurse Practitioner Program at New York University College of Nursing; developed the BS in Holistic Nursing Program at Pacific College of Oriental Medicine; and is past President, American Holistic Nurses Association.

Philosophy, Mission, and Goals

In 2006, holistic nursing was recognized by the American Nurses Association (ANA) as an official nursing specialty with a defined scope and standards of practice within the discipline of nursing. Holistic nursing emanates from five core values summarizing the ideals and principles of the specialty. These core values are:

- Holistic Philosophy, Theory, Ethics
- Holistic Caring Process
- Holistic Communication, Therapeutic Healing Environment, and Cultural Diversity
- Holistic Education and Research
- Holistic Nurse Self-Reflection and Self-Care

Holistic Nursing:

- embraces all nursing which has enhancement of healing the whole person across the life-span and the health-illness continuum as its goal

- recognizes the interrelationship of the unified bio-psychosocial-cultural-spiritual-energetic-environmental dimensions of the person
- focuses on protecting, promoting, and optimizing health and wellness, assisting healing, preventing illness and injury, alleviating suffering, and supporting people to find meaning, peace, comfort, harmony, and balance through the diagnosis and treatment of human response
- views the nurse as an instrument of healing and a facilitator in the healing process
- honors the individual's subjective experience about health, illness, health beliefs, and values
- uses the caring-healing relationship and therapeutic partnership with individuals, families, and communities
- draws on nursing knowledge, theories, research, expertise, intuition, and creativity
- incorporates the roles of clinician, educator, consultant, coach, partner, role model, and advocate
- encourages peer review of professional practice in various clinical settings and utilizes knowledge of current professional standards, laws, and regulations governing nursing practice
- collaborates and partners with all constituencies in the health process including the person receiving care, family, significant others, community, peers, and other disciplines, using principles and skills of cooperation, alliance, consensus, and respect, and honoring the contributions of all.
- focuses on integrating self-reflection, self-care, and self-responsibility in personal/professional life
- emphasizes awareness of the interconnectedness of self, others, nature, and God/Life/Spirit/Universal Force

Characteristics and Data

There are 3.1 million nurses in the US, of which 6,000–10,000 identify as holistic nurses and 5,750 are members of the American Holistic Nurses Association (AHNA).

Scope of Practice

- draw on nursing knowledge, theories of wholeness, research and evidence-based practice, expertise, caring, and intuition to become therapeutic partners with clients and significant others in a mutually evolving process toward healing, balance, and wholeness
- integrate holistic, alternative, complementary and integrative modalities including, for example, relaxation, meditation, guided imagery, breath work, biofeedback, aroma and music therapies, touch therapies, acupressure, herbal remedies and natural supplements, homeopathy, reflexology, Reiki, journaling, exercise, stress management, nutrition, self-care processes, and prayer with conventional nursing interventions
- conduct holistic assessments of physical, functional, psychosocial, emotional, mental, sexual, cultural, age-related, spiritual, beliefs/values/preferences, family issues, lifestyle patterns, environmental, and energy field status
- select appropriate interventions in the context of the client's total needs and evaluate care in partnership with the client
- assist clients to explore self-awareness, spirituality, growth, and personal transformation in healing
- work to alleviate clients' signs and symptoms while empowering clients to access their own natural healing capacities
- concentrate on the underlying meanings of symptoms and changes in the client's life patterns

- provide comprehensive health counseling, education and coaching, health promotion, disease prevention, and risk reduction
- guide clients/families between the conventional allopathic medical system and complementary/alternative/integrative therapies and systems
- collaborate with and refer to other healthcare providers/resources as necessary
- advocate to provide access to and equitable distribution of healthcare resources, and to transform the healthcare system to a more caring culture
- participate in building an ecosystem that sustains the wellbeing of the environment and the health of people, communities, and the planet
- practice in numerous settings including acute, ambulatory, community, and home care, private practice, wellness/complementary/integrative care centers, women's health centers, schools, employee/student health, psychiatric mental health facilities, rehabilitation centers, correctional facilities, Telehealth/cyber care services, and colleges/universities.
- Holistic nurses with advanced education can become advanced practice nurses, faculty, administrators, and researchers

Education

Nursing Programs in the US[1]
243 Doctoral programs (PhD, EdD, DNS, DNP)
476 Master's programs (MA, MS, MSN, MEd)
677 Baccalaureate programs (BS, BSN)
1060 Associate programs (AD, ADN)

American Holistic Nurses Certification Corporation (AHNCC)-Endorsed Academic Programs in Holistic Nursing
5 Graduate; 10 Undergraduate

Nursing Accrediting Bodies
Commission on Collegiate Nursing Education (CCNE)
Accreditation Commission for Education in Nursing (ACEN)

Certifying Organizations for Holistic Nursing
American Holistic Nurses Certification Corporation (AHNCC)

Regulation and Certification

- National Licensure Exam (NCLEX) for all registered nurses through the National Council of State Boards of Nursing
- 25 (47%) State Boards of Nursing have a formal policy, position, or inclusion of holistic/complementary/alternative therapies in the scope of practice of nurses
- National Board Certification in Holistic Nursing at the basic (HN-BC – diploma and associate degree), (HNB-BC – baccalaureate degree) or advanced (AHN-BC- master's and doctoral) levels through the American Holistic Nurses Certification Corporation (AHNCC)
- American Holistic Nurses Association (AHNA) and American Nurses Association (ANA) co-published *Holistic Nursing: Scope and Standards of Practice, 2nd ed. (2013).*

Research

The following are examples of the types of research being done by holistic nurse researchers.

Quantitative
Outcome measures of various holistic therapies, e.g., therapeutic touch, prayer, presence, relaxation, aromatherapy; instrument development to measure holistic phenomena; caring behaviors and dimensions; spirituality; self-transcendence; cultural competence; client responses to holistic interventions in health/illness/wellness.

Qualitative
Explorations of clients' lived experiences with various health/illness/life phenomena; theory development in healing, caring, wellbeing, intentionality, social and cultural constructions, empowerment, health decision making; healing relationships and environments; health and wellness promotion; etc.

Challenges and Opportunities 2012–2018

Education

- integration of holistic philosophy, content, and practices into nursing curricula nationally and staff development programs
- recognition, support, and legitimization of holistic integrative nursing practice in accreditation, regulation, licensure, and credentialing processes

Research

- identification and description of outcomes of holistic therapies/interventions, relationships, and environments
- focus on whole systems research and wellness, health promotion, and illness prevention
- funding nurses for CAM and wholeness research

- dissemination of nursing research findings to broader audiences including other health disciplines and public media

Practice

- influence and change the healthcare system to a more holistic, humanistic orientation
- development of caring cultures within healthcare delivery models and systems
- collaboration with diverse healthcare disciplines to advance holistic health care
- improvement of the nursing shortage through incorporation of self-care and stress-management practices for nurses and improvement of healthcare environments

Policy

- coverage and reimbursement for holistic nursing practices and services
- education of the public about the array of healthcare alternatives and providers
- increase focus on wellness, health promotion, access, and affordability of health care to all populations
- care of the environment and the planet

Resources

Organizations

American Holistic Nurses Association (AHNA), founded 1982, the definitive voice of holistic nursing, provides vision, direction, and leadership in developing and advancing holistic philosophy, principles, standards, and guidelines for practice, education, and research. AHNA's mission is "to advance holistic nursing through community building, advocacy, research, and education". AHNA is committed to promoting wholeness and wellness in individuals, families, communities, nurses themselves, the nursing profession, and the environment. Through its various activities, AHNA integrates the art and science of nursing in the profession; unites nurses in healing; focuses on health, preventive education, and the integration of allopathic and CAM/integrative caring healing modalities; honors individual excellence in the advancement of holistic nursing; and influences policy for positive change in the healthcare system. Phone: 800-278-2462, Email: info@ahna.org; www.ahna.org.

American Nurses Association (ANA) represents the interests of the nation's 3.1 million registered nurses (RNs) through its 55 constituent member associations, and its 29 specialty nursing and workforce advocacy organizations that currently connect to ANA as affiliates. The ANA advances the nursing profession by fostering high standards of nursing practice, promoting the rights of nurses in the workplace, projecting a positive and realistic view of nursing, and by lobbying Congress and regulatory agencies on healthcare issues affecting nurses and the public. http://www.nursingworld.org.

Citations
1. American Association of Colleges of Nursing. Institutions offering doctoral programs in nursing and degrees conferred. Fall 2012. www.aacn.nche.edu/research-data/doc.pdf
2. American Holistic Nurses Association/American Nurses Association. Holistic Nursing: Scope and Standards of Practice, 2nd ed. Silver Spring, MD: Nursesbooks.org. 2013

Homeopathy

First Edition (2009) Author: Todd Rowe, MD, MD(H), CCH, DHt

Second Edition (2013) Editor: Heidi Schor, CCH, LMP, CC

First Edition (2009) Partner Organization: American Medical College of Homeopathy (AMCH)

Second Edition (2013) Partner Organization: Accreditation Commission for Homeopathic Education in North America (ACHENA)

About the Authors/Editors: Todd Rowe is Founder and President of the American Medical College of Homeopathy. He is President of the Arizona Board of Homeopathic and Integrated Medicine Examiners, has served on the board of directors for the Council for Homeopathic Certification and is a past President of the National Center for Homeopathy. Heidi Schor is immediate past President of the Accreditation Commission for Homeopathic Education in North America, a board member of the Council for Homeopathic Certification, past President of the Washington State Homeopathy Association, former faculty for Bastyr University, Seattle School of Homeopathy and the Homeopathic Academy of Southern California, and served as Founder and Director of the Homeopathy Community Clinic of Kirkland, a low income student-learning clinic.

Philosophy, Mission, Goals

Homeopathic medicine is holistic, informed by science, safe to use, and inexpensive. Homeopathy can be effective in both acute and chronic disease and can be useful in many emergency situations as well. It is a system of medicine that is unique and distinct from other systems, disciplines, or modalities such as biomedicine, naturopathic medicine, herbal medicine, acupuncture, nutritional medicine, and mind-body medicine. Homeopathy uses minute doses of natural substances to activate the body's self-regulatory healing mechanisms. It was founded by Samuel Hahnemann over 200 years ago, although the principles on which it is based have been utilized in healing for thousands of years. Since its inception, homeopathy has been used by people from all walks of life, all ages, and in countries all over the world.

Characteristics and Data

Homeopathic medicine is the second most common form of alternative medicine in the world today and the most common in some higher income countries.[1] Homeopathic medicine was first introduced into the United States in 1825 where it flourished until around 1900 when the field began to meet opposition from conventional medicine.

At that time, there were 22 homeopathic medical colleges and 20 percent of physicians used homeopathic medicine. The number of current homeopathic practitioners in this country is estimated at 8500.[2] There are four subgroups practicing homeopathy in some form in the United States. Standards, regulation, licensure, and educational accreditation vary widely between these groups.

- *Lay homeopaths:* No training standards, testing or recognition. Member organization is National Center for Homeopathy (NCH).
- *Professional homeopaths:* Not licensed as healthcare professionals; typically 850-1200 hours of training; Council on Homeopathic Certification (CHC) tests; no external recognition, accreditation, or licensure; member organization is North American Society of Homeopaths (NASH). Health freedom legislation covers these practitioners in some states.
- *Registered homeopathic medical assistant:* 300 hours of training required by state registration boards in Arizona and Nevada; MD or DO supervision required.
- *Licensed healthcare professionals:* Training ranges from 200–300 hours to 1000 hours, depending on the field and the level of specialization.

Scope of Practice

The scope of practice varies considerably from state to state and is defined by various licensing agencies as well as, in some states, health freedom legislation. Homeopathic licensing boards exist in Arizona, Connecticut, and Nevada for MDs/DOs; in 15 states a section of naturopathic medical board exams is on homeopathy.

Integration Activities

Homeopathic schools offer services in many off-site clinics. Examples of clinical areas of practice include fertility, acute care, incarcerated adolescents, low-income Hispanic population, substance abuse, homeless/free clinic, urgent care centers, and nursing homes. Homeopathic practitioners are sought out to provide expertise in the field of complementary medicine, including policy development, medical training, medical research, and clinical applications of natural therapies. Homeopaths Without Borders offers homeopathy volunteer services to individuals living in regions facing crisis conditions. Combined degree programs are offered in homeopathic medicine and acupuncture.

Education

There are currently thirty homeopathic schools in the US, fifteen of which are listed in the school directory of the North American Network of Homeopathic Educators (NANHE) and three of which are recognized by the Accreditation Commission for Homeopathic Education in North America (ACHENA), the recognition body for homeopathic schools. ACHENA is actively pursuing recognition by the US Department of Education. Naturopathic medicine is the only healthcare profession which requires homeopathic training based on Department of Education-recognized accrediting standards for all of its practitioners. Some conventional medical schools and residencies offer electives in homeopathic medicine.

Regulation and Certification

Laws regulating the practice of homeopathy vary from state to state. Usually it can be practiced legally by those whose license entitles them to practice a healthcare profession or medicine (see Scope of Practice above). Health freedom laws in some states allow the practice of homeopathy by non-licensed professionals. The Food and Drug Administration regulates the manufacture and sale of homeopathic medicines in the US. The Homeopathic Pharmacopoeia of the United States was written into federal law in 1938 under the Federal Food, Drug, and Cosmetic Act, making the manufacture and sale of homeopathic medicines legal in this country.[3] Since homeopathic remedies are sold over the counter, people in all states are free to use them for self-care at home.

Graduates of schools are eligible to test for national certification through various certifying organizations, the largest of which is the Council for Homeopathic Certification (CHC). Though the conventional medical doctor (MD) and chiropractic professions each have specialty certification exams, the naturopathic doctors use the CHC exam.

Research

The homeopathic profession has been conducting research for over 200 years. Several hundred studies have been published in recent years in clinical homeopathic research. A leading organization for homeopathic research is the Society for the Establishment of Research in Classical Homeopathy (SERCH). Homeopathic research focuses on five areas:

- basic sciences research
- clinical sciences research
- educational research
- homeopathic research

- practice-based research

Challenges and Opportunities

Key challenges 2012–2015

- increasing public visibility of homeopathic medicine
- establishing widespread legality of practice
- need for expansion of homeopathic research
- opposition to regulation of the profession
- continued availability of Homeopathic Medicines
- continued improvement of educational standards for the profession
- requirement of accredited education for national certification

Key opportunities 2012–2015

- potential for the development of bridge programs in homeopathic medicine for other healthcare professionals
- increased demand for homeopathic services
- increased participation in Integrative Medicine field
- growth and professionalization of homeopathic schools
- active health freedom initiatives in multiple states
- establishment of national homeopathic medical school
- CHC attainment of Institute for Credentialing Excellence accreditation
- ACHENA attainment of Department (Secretary) of Education recognition
- inclusion in the Affordable Care Act as part of national healthcare workforce as integrative healthcare practitioners

Resources

Organizations
The following are the principal national homeopathic organizations:

- Academy of Veterinary Homeopathy (AVH)
  www.theavh.org
- Accreditation Commission for Homeopathic Education in North America (ACHENA) www.achena.org
- American Association of Homeopathic Pharmacists (AAHP)
  www.homeopathicpharmacy.org
- American Institute of Homeopathy (AIH)
  www.homeopathyusa.org
- Council for Homeopathic Certification (CHC)
  www.homeopathicdirectory.com
- Homeopathic Academy of Naturopathic Physicians (HANP)
  www.hanp.net
- Homeopathic Nurses Association (HNA)
  www.nursehomeopaths.org
- Homeopathic Pharmacopoeia of the United States
  www.hpus.com/overview.php
- Homeopaths Without Borders (HWB)
  www.homeopathswithoutborders-na.org
- National Center for Homeopathy (NCH)
  www.nationalcenterforhomeopathy.org
- North American Network of Homeopathic Educators (NANHE) www.homeopathyeducation.org
- North American Society of Homeopaths (NASH)
  www.homeopathy.org

NCH is the national consumer association, NASH is the national association for homeopathic practitioners. ACHENA is the association for recognition of schools. The CHC, HANP and ABHT are the main national certification bodies for homeopathic practitioners.

Citations

1. Ong CK, Bodeker G, Grundy C, et al. *World Health Organization Global Atlas of Traditional Complementary and Alternative Medicine (Map Volume)*. Kobe, Japan. World Health Organization, Centre for Health Development, 2005:61
2. National Homeopathic Practitioner Survey. January 2007. http://www.amcofh.org/research/community , Accessed November 17, 2009.
3. Homeopathic Pharmacopoeia of the United States www.hpus.com/overview.php. Accessed November 17, 2009.

Integrative Medicine

First Edition (2009) Author: Victor S. Sierpina, MD

Second Edition (2013) Editor: Benjamin Kligler, MD

Partner Organization: Consortium of Academic Health Centers
for Integrative Medicine (CAHCIM)

About the Authors/Editors: Sierpina is the WD and Laura Nell
Nicholson Professor of Integrative Medicine and Professor of
Family Medicine at the University of Texas Medical Branch. He
is the former Chair of the Consortium of Academic Health
Centers for Integrative Medicine. Kligler is an Associate
Professor of Family Medicine at Albert Einstein College of
Medicine and Vice Chair of the Department of Integrative
Medicine at Beth Israel Medical Center. He is the current Chair
of the Consortium of Academic Health Centers for Integrative
Medicine

Characteristics and Data

Integrative medicine is a movement to transform medicine led
by a growing set of leaders from conventional academic health
centers in association with community-based practitioners from
diverse disciplines. The lead organization in that movement is
the Consortium of Academic Health Centers for Integrative
Medicine (CAHCIM).

CAHCIM is composed of 54 North American academic health
centers and health systems which have active programs in at
least two of the three areas of education, clinical care, and
research in integrative medicine and have support at the Dean's
level or above. The Consortium defines integrative medicine as
"...the practice of medicine that reaffirms the importance of the

relationship between practitioner and patient, focuses on the whole person, is informed by evidence, and makes use of all appropriate therapeutic approaches, healthcare professionals and disciplines to achieve optimal health and healing."

The Consortium started with eight schools in 1999 and, with the generous support of the Bravewell Philanthropic Collaborative for infrastructure, has grown rapidly expanding to 23 schools by 2003, 36 in 2006, and 52 in 2012. For a complete list of member schools and other Consortium information, see the CACHIM website listed in the Resources section.

The mission of the Consortium is "to advance the principles and practices of integrative healthcare within academic institutions." The Consortium provides its institutional membership with a community of support for their academic missions and a collective voice for influencing change. The goals of the Consortium are:

- stimulate changes in medical education that facilitate the adoption of integrative medicine curricula
- develop and promote policies and practices that enhance access to integrative models of care
- undertake collaborative projects to promote and conduct research and disseminate the findings
- inform national policy that advances integrative medicine through research, clinical, and education initiatives

With the voice of over one third of US and Canadian medical schools, the Consortium is poised to continue to advocate within academia for a relevant role for integrative medicine, across all institutional missions. The Consortium and the Academic Consortium for Complementary and Alternative Health Care (ACCAHC) have collaborated in planning two major international research conferences in 2009 and 2012, and co-sponsored a conference on education in integrative health care in October 2012.

Leadership and Working Groups

The Consortium leadership is composed of an Executive Committee (EC) comprised of the Chair, Vice-Chair, Treasurer, Secretary, and the immediate past Chair, as well as four at-large members. Chairs of the Working Groups also sit on the EC, as does the chair of the Membership Committee. The EC meets monthly by phone and in person twice annually. Officers are elected by a Steering Committee (SC), which consists of one representative from each of the 54 member institutions. The SC meets in person twice annually. The Consortium is governed by by-laws and is an IRS-recognized nonprofit 501(c)(3) organization.

Education Working Group
The Education Working Group (EWG) is continuously involved in fostering projects in integrative medical education. Previous projects have included creating a curriculum guide containing exemplars of integrative medicine teaching, authoring an article on proposed competencies for a medical school curriculum in integrative medicine, and coordinating a grant to foster mind-body training and curriculum among the Consortium schools. There is now a Student Leadership Group which has emerged from the EWG, with student representatives from member institutions, and there will soon be a resident/fellow group as well to serve students after they move on to postgraduate training. Other current EWG projects include an effort to collect and assess existing tools for evaluating outcomes of educational interventions in integrative medicine, and another to collect currently available online teaching resources from member institutions for more effective dissemination.

A major accomplishment of the EWG has been the "Leadership and Education Program in Integrative Medicine (LEAPS into IM)" project, now entering its fourth year. This program, a collaboration with the American Medical Student Association and the Kripalu Yoga Center, offers a one-week, immersion-style training program in integrative medicine leadership to 30 medical students annually from across North America.

Clinical Working Group

The Clinical Working Group (CWG) examines clinical initiatives across the member schools and shares expertise on clinical care, clinical models, and business issues, as well as fostering clinical research. In addition to the core working group, the CWG has created collaborations with several subgroups in specific clinical areas, including integrative oncology, pediatrics, pain medicine, cardiology, integrative mental health, and financial/business models. These groups meet on a regular basis via telephone, offering talks to members from leaders in each particular area. The calls are open to all Consortium members and are recorded and stored on the website for reference. The CWG was involved in planning content for the 2012 Annual Meeting on clinical models of care across the Consortium institutions.

Research Working Group

The Research Working Group (RWG) has created a network of researchers from Consortium institutions to share their expertise and patient populations for recruitment in studies in integrative medicine. The RWG recently created four sub-groups to allow researchers with specific areas of interest to come together more readily for possible collaboration and collective learning. These groups, which will meet regularly by conference call, are in the areas of cancer symptomology, workplace wellness, models of integrative healthcare and personalized healthcare.

The RWG also plays a major role in planning the Consortium's Research Conference, for which members serve as contributors and scientific reviewers. In addition, in 2011, the RWG organized a very successful one-day research meeting following the Consortium Annual Meeting, which was attended by over 100 researchers and NCCAM leadership.

Policy Working Group

The Policy Working Group (PWG) of the Consortium informs the Steering Committee of policy level issues that could influence the mission of the Consortium and provides information to policy makers regarding the Consortium and integrative medicine, including issues relevant to integrative medicine education, research and clinical practice. A priority of the PWG is to continue to monitor development and implementation of federal legislation and advise the Consortium of opportunities related to Integrative Medicine. The PWG also coordinates requests for testimony at Congressional briefings or other hearings pertaining to government policy.

Fellowship Task Force

This ad hoc group was commissioned by the Consortium leadership in 2011 to develop a consensus set of competencies for fellowship level education in integrative medicine. The Task Force has systematically reviewed existing competencies and characteristics of currently available US integrative medicine clinical fellowships, and has drafted a set of standard integrative medicine clinical fellowship competencies. These draft competencies are currently being revised in preparation for publication. We are hopeful that this project will provide a blueprint for future formal accreditation of fellowship programs in integrative medicine.

Challenges and Opportunities 2012-2015

While the future of integrative medicine education is bright, some major challenges and opportunities lie ahead for us to prepare the next generation of integrative medicine practitioners. Among these are:

- faculty development in both conventional and CAM schools regarding the content of the field, teaching methods, and research expertise
- interprofessional collaboration in education, research, clinical care, and healthcare policy between conventional and CAM disciplines
- providing learners with critical thinking skills, access to reliable resources, and role modeling of integrative practice in consistent, credible, and relevant ways
- outcomes-based practice and optimal care pathways informed by research
- changes in health insurance reimbursement and policy to provide broader access to integrative medicine for under-served patients

Resources

Organizations and Websites

- Consortium of Academic Health Centers for Integrative Medicine (CAHCIM)
  www.imconsortium.org

Conferences

*   Research conference: International Congress on Integrative Medicine and Health (May 2006, May 2009, May 2012). Next conference scheduled for May 2014 in Miami.

Bibliography
Consortium-endorsed:
Kligler B, Maizes V, Schachter S.Core competencies in integrative medicine for medical school curricula: a proposal. *Acad Med*. 2004; 79:521-531.

Featured articles:
Academic Medicine issue with 10 articles on CAM educational programs in medical and nursing schools, with major involvement by multiple Consortium programs. *Acad Med*. 2007; 82.

Yoga Therapy

First Edition (2009) Author and Second Edition (2013) Editor:
John Kepner, MA, MBA

Partner Organization: International Association of Yoga
Therapists (IAYT)

About the Author/Editor: Kepner is the Executive Director of the
International Association of Yoga Therapists.

Philosophy, Mission, and Goals

The International Association of Yoga Therapists (IAYT)
supports research and education in yoga and serves as a
professional organization for yoga teachers and yoga therapists
worldwide. Our mission is to establish yoga as a recognized and
respected therapy. IAYT published a definition of the field in
2012 as part of our standards effort. In brief:

> Yoga therapy is the process of empowering individuals to
> progress towards improved health and wellbeing through
> the application of the teachings and practices of yoga....
> The practices of yoga traditionally include, but are not
> limited to, *asana* (postures), *pranayama* (breath control),
> meditation, mantra, chanting, *mudra*, ritual and a
> disciplined lifestyle.[1]

In contemporary usage, the distinction between yoga and
yoga therapy is not clear cut and depends, in part, upon the
language used by different yoga lineages and methods and the
context. In general, however, yoga therapy is an orientation to
the practice of yoga that is focused on healing, as opposed to

fitness, wellness, or spiritual support. It is often taught privately or for special populations.

Characteristics and Data

Yoga is one of the most popular complementary and alternative medicine (CAM) therapies - used by approximately 6.1% of the adult population. The use of yoga, as well as meditation and deep breathing (both considered yogic practices) all appear to be on the rise.

Figure 2.1

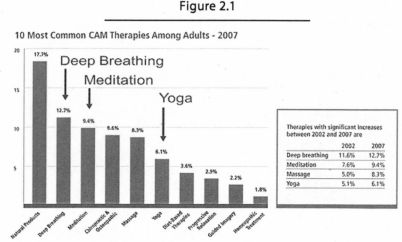

From: Complementary and Alternative Medicine Use Among Adults and Children: United States, 2007, Barnes PM, Bloom B, Nahin R. CDC National Health Statistics Report #12, 2008.

The practice of yoga therapy is unregulated, and currently individuals are free to describe themselves as yoga therapists. The field is growing rapidly. IAYT has over 3,000 individual members and 100 member schools offering certification as a yoga therapist, up from approximately 640 individual members and 5 member schools in 2003.

Scope of Practice

The scope of practice varies by method and philosophy, but most commonly can be described as the management of structural aches and pains, chronic disease, emotional imbalance, stress, and spiritual challenges using the student's own body, breath, and mind.

Integration Activities

IAYT has been a member of ACCAHC since 2006 and is a founding member of the Traditional World Medicine Group within ACCAHC. The association enjoys a close working relationship with our sister organizations, the National Ayurvedic Medical Association (NAMA), and the Yoga Alliance (YA). Yoga therapists and therapeutic yoga classes are beginning to appear in hospitals, medical clinics and mental health practices. There appears to be more and more integration with other practices, especially physical therapy, mental health services and chiropractic. Almost 40% of the IAYT membership holds a healthcare license in other fields, with the top three being mental health, physical therapy and massage.

Education

The training of yoga therapists is most commonly provided by private schools and/or supported by traditional yoga lineages. At least two certification courses are currently being offered by major universities in North America. At least three master's and one PhD program with a concentration in yoga therapy appear to be currently available or announced for 2012. One actual Master of Science degree in yoga therapy from an accredited university was announced in 2013.

IAYT published voluntary educational standards for the training of yoga therapists in 2012 and will begin accrediting training programs in 2013. Credentialing practitioners is expected to begin after the accreditation process is firmly established. These standards are focused on competencies and require considerably more training than the majority of current training programs. The announcement of these standards appears to have spurred additional interest by institutions of higher education.

Regulation and Certification

Yoga therapy is an unregulated discipline. In the US, the profession is represented by the International Association of Yoga Therapists (IAYT). IAYT is currently spearheading an effort to improve the professional status of the field, including implementing training standards for yoga therapists, accrediting yoga therapy training programs, and, it is expected, eventually credentialing Yoga therapists.

Currently, individual schools provide certificates to graduates. Credentialing by IAYT is on the horizon after the program accreditation process is established. To date, the field has resisted outside regulation and licensing but, instead, has supported self-regulation through a voluntary standards process.

Research

The National Institutes of Health supports basic research into yoga. Yoga is classified as mind-body medicine by the National Center for Complementary and Alternative Medicine (NCCAM).

IAYT is actively involved with promoting research in the field of yoga therapy. The association successfully launched a standalone academic yoga research conference, the Symposium on Yoga Research (SYR) in 2010, with over 200 attendees (three times expectations). The second conference, SYR 2011, was awarded an NCCAM R-13 scientific conference support grant. See the Symposium on Yoga Therapy and Research website in the Resources section for current information on IAYT professional research conferences.

Members of the association are active participants in the bi-annual integrative medicine research conference sponsored by the Consortium of Academic Health Centers for Integrative Medicine (CAHCIM). Each of the first three conferences featured a symposium on yoga research.

IAYT's journal, the *International Journal of Yoga Therapy* (IJYT), was accepted by Pub-Med in 2011.

Challenges and Opportunities

Key challenges 2012-2015

- Implementing the educational standards, including accrediting yoga therapist training programs.
- Credentialing yoga therapists, including grandparenting current practitioners.
- Developing sound models for the appropriate integration of this unlicensed field with conventional, complementary and alternative healthcare fields.
- Financing the training required by the new educational standards, given the still emerging nature of the field and the limited opportunities for economically sustainable employment.

Key opportunities 2012-2015

The practice of yoga is a low-cost, effective practice for promoting wellness and supporting healing at many levels. A key element for yoga is that, once trained, individuals can practice without professional supervision.

The practice has grown tremendously over the past ten years, in part because the practice is so adaptable to the young and the old, the fit and the not-so-fit, the exercise enthusiast and the spiritual seeker. It is an especially suitable wellness practice for an aging population.

Given the growth and popularity of the field in the West in the past 10 years, the emerging research supporting its many healing properties, the growing acceptance of complementary and alternative therapies in general, and the high cost of conventional care, there are many opportunities to develop the field as a stand-alone healing practice or in conjunction with conventional and CAM healthcare fields.

Since yoga is not a licensed practice, the practice is not constrained by insurance reimbursement practices.

Organizations and Websites

The International Association of Yoga Therapists (IAYT) is a professional association with approximately 3,000 individual members and 100 member schools in almost 50 countries (as of 2013). IAYT supports research and education in yoga and serves as a professional organization for yoga teachers and yoga therapists worldwide. Our mission is to establish yoga as a recognized and respected therapy. IAYT publishes the *International Journal of Yoga Therapy,* a peer-reviewed journal and *Yoga Therapy Today,* a practitioner's magazine; sponsors an annual Symposium on Yoga Therapy and Research (SYTAR), and an annual research conference, the Symposium on Yoga Research (SYR) and provides a search function for the public to find members in their area.

- International Association of Yoga Therapists (IAYT) www.iayt.org

- Symposium for Yoga Therapy and Research
  www.sytar.org

Citations

1. International Association of Yoga Therapists (IAYT).
Educational Standards for the Training of Yoga Therapists.
July 1, 2012.
http://www.iayt.org/development_Vx2/IAYT_Standards_7
%201%2012%20.pdf

Appendices

1. About the Academic Consortium for Complementary and Alternative Health Care (ACCAHC)

2. Accomplishments: January 2005-January 2013

3. Toward Collaboration between the Disciplines: An ACCAHC Perspective

4. ACCAHCcronyms

Appendix 1

About the Academic Consortium for Complementary and Alternative Health Care (ACCAHC)

[We] envision a healthcare system that is multidisciplinary and enhances competence, mutual respect and collaboration across all healthcare disciplines. This system will deliver effective care that is patient centered, focused on health creation and healing, and readily accessible to all populations.

—Vision Statement, Academic Consortium for
Complementary and Alternative
Health Care (ACCAHC)
Endorsed March 2005; Amended July 2012

The idea of an organization like ACCAHC initially originated in a series of meetings beginning in the late 1990s: the Integrative Medicine Industry Leadership Summits. Additional impetus was provided by the National Policy Dialogue to Advance Integrated Health Care (2001). In each instance, leading educators in integrative practice recognized the value of creating a vehicle for ongoing interdisciplinary dialogue and action.

These seeds began to take formal shape in 2003 as a project of the Integrated Healthcare Policy Consortium (www.ihpc.info). The project evolved until the end of 2006 under the direction of Pamela Snider, ND, founding executive director of ACCAHC, and Reed Phillips, DC, PhD, founding ACCAHC chair.

ACCAHC's key purpose was to bring educator leaders of the five licensed complementary and alternative healthcare professions into one room so they could jointly articulate their shared issues and concerns about the evolving integrative medicine dialogue. ACCAHC allowed educators to create one voice as they joined with their academic counterparts in conventional medicine in the 2005 National Education Dialogue to Advance Integrated Health Care: *Creating Common Ground* (NED), which was led (under IHPC's auspices) by John Weeks, Snider's successor as ACCAHC executive director.

Once convened in the same room, however, ACCAHC's educator leaders realized there were many additional values in ongoing dialogue and collaboration. These educator leaders determined in mid-2006 that ACCAHC should seek to become an independent entity to imbed this movement as a continuing part of the landscape. ACCAHC was subsequently incorporated as a nonprofit corporation, later obtaining Internal Revenue Service recognition as a tax-exempt 501(c)(3) organization. Table A.1 lists the organizational members of ACCAHC, each of which pays dues based on the size of the organization.

ACCAHC's core members are those councils of colleges and schools, US Department of Education-recognized accrediting agencies, and nationally recognized certification and testing organizations from the licensed complementary and alternative healthcare disciplines that choose to join. Current ACCAHC core membership includes 13 of these 15 organizations. In addition, schools, colleges and universities affiliated with roughly two dozen of these accredited programs have chosen to join ACCAHC as individual institutional members.

As we developed our structure, we also decided not to close the door on professions that are committed to greater self-regulation but have not yet met the benchmarks of the licensed disciplines. Concerns were voiced that without such a membership provision ACCAHC would recapitulate the pattern of exclusion that has challenged the maturation processes of the currently licensed professions. ACCAHC developed distinct membership categories for two new groups, traditional world medicines (TWM) and emerging professions (EP). At this time, two TWM organizations related to Yoga therapy and Ayurvedic medicine and two EP organizations related to homeopathic medicine had applications for membership accepted, as indicated in Table A.1.

Table A.1

ACCAHC Organizational Members as of March 2013

Member Type	Member Organizations
Councils of Colleges and Schools	Alliance for Massage Therapy Education www.afmte.org Association of Accredited Naturopathic Medical Colleges www.aanmc.org Association of Chiropractic Colleges www.chirocolleges.org Council of Colleges of Acupuncture and Oriental Medicine www.ccaom.org
Accrediting Agencies	Accreditation Commission for Acupuncture and Oriental Medicine www.acaom.org Commission on Massage Therapy Accreditation www.comta.org Council on Chiropractic Education www.cce-usa.org Council on Naturopathic Medical Education www.cnme.org Midwifery Education Accreditation Council www.meacschools.org
Testing and Certification Organizations	National Board of Chiropractic Examiners www.nbce.org National Certification Board for Therapeutic Massage and Bodywork www.ncbtmb.org National Certification Commission for Acupuncture and Oriental Medicine www.nccaom.org North American Board of Naturopathic Examiners www.nabne.org
Traditional World Medicines and Emerging Professions*	Accreditation Commission for Homeopathic Education in North America http://www.achena.org/ Council for Homeopathic Certification http://www.homeopathicdirectory.com/ International Association of Yoga Therapists www.iayt.org National Ayurvedic Medical Association www.ayurveda-nama.org

*The Traditional World Medicines and Emerging Professions categories were created to allow participation of professions engaged in the educational and professional regulatory efforts described earlier

As mentioned above, ACCAHC reflects collaboration with a growing list of donors. Membership funds roughly a fifth of basic operations with the remainder of operational funding and all major projects dependent on foundations, corporate, and individual philanthropy. We also plan to create additional funding through the provision of resources that assist us in fulfilling on our vision. Donors that have assisted ACCAHC significantly in advancing our mission include ACCAHC's founding philanthropic partner Lucy Gonda, the Westreich Foundation, the Lia Fund, the NCMIC Foundation, the Leo S. Guthman Fund, Standard Process, Inc., Bastyr University, Life University, Andrew Weil, MD, and the Consortium of Academic Health Centers for Integrative Medicine. The Institute for Alternative Futures provided us with significant in-kind support. Finally, ACCAHC is deeply grateful for the visionary leaders of the Integrated Healthcare Policy Consortium out of whose multidisciplinary, collaborative work ACCAHC was born and then nurtured in its first years.

ACCAHC's projects are led through the Board of Directors, an Executive Committee, and three working groups, one each in the areas of research, education, and clinical care. The projects are diverse, and each includes a focus on supporting inter-professional education, research, and clinical practice. Among current or recent projects are: developing the ACCAHC *Competencies for Optimal Practice in Integrated Environments*; engaging productive dialogues with the NIH National Center for Complementary and Alternative Medicine over that agency's 2011-2015 strategic plan and the Patient Centered Outcomes Research Institute in its initial priorities; engaging the emerging movement for interprofessional practice and education via diverse forums and relationship building including participation in the new Center for Interprofessional Practice and Education; completing a project with UCLA Center for Health Policy senior fellow Michael Goldstein on the role of licensed integrative health disciplines in meeting the nation's primary care needs; and co-sponsoring the hugely successful International Congress for Educators in Complementary and Integrative Medicine at Georgetown University in October 2012 with our colleagues in the Consortium of Academic Health Centers for Integrative Medicine.

This level of engagement stimulated significant interest in ACCAHC playing a role in skill development for leadership to bring the values, practices and disciplines associated with integrative health care into the broader dialogue over the future of medicine and health. The Board of Directors established a Task Force on Leadership Development in 2011 to respond to this need. One focus is on training individuals to serve as "ambassadors" or external affairs representatives to diverse healthcare initiatives. These individuals act as translators. They advance ACCAHC's values of whole person, health-focused care in education, clinical services, research and policy. At the same time, they help communicate and connect significant trends and developments in medicine and health to the ACCAHC community.

ACCAHC representatives are currently serving such organizations and initiatives as: the Pain Action Alliance to Implement a National Strategy; the Health Resources and Services Administration (HRSA)-funded National Coordinating Center for Integrative Medicine via the American College of Preventive Medicine; the biennial Collaboration Across Borders conference on interprofessional practice and education; the National Quality Forum; and are engaged in dialogue with the leadership of the emerging National Center for Interprofessional Practice and Education.

Of particular note is ACCAHC's work with three significant projects of the Institute of Medicine of the National Academy of Sciences. The first ACCAHC engagement was to support consumer interest in distinctly licensed complementary and alternative healthcare practitioners during planning and execution of the Institute of Medicine's (IOM) February 25–27, 2009 National Summit on Integrative Medicine and the Health of the Public. Conversations between leaders of ACCAHC and representatives of the IOM led to the appointment of ACCAHC chair Elizabeth Goldblatt, PhD, MPA/HA to the IOM's planning committee. Goldblatt's service included helping form valuable links between the conventionally trained practitioners and educators who were her colleagues on the committee and key professionals in the licensed complementary healthcare disciplines. The rich resources created through the Summit are available through the IOM's site http://www.iom.edu/?ID=52555.

Subsequently, the IOM responded affirmatively to a suggestion from ACCAHC that a committee that was formed to develop a national strategy for pain research, education and care should include at least one representative from the licensed complementary healthcare disciplines. We made the case that many consumers use chiropractors, acupuncture and Oriental medicine practitioner, massage therapist, Yoga therapists and naturopathic doctors for pain-related conditions. An ACCAHC nominee, Rick Marinelli, ND, LAc, a past president of the American Academic of Pain Management, was selected. We were pleased to see that the IOM report, "Relieving Pain in America: A Blueprint for Transforming Prevention, Care, Education, and Research," promoted an integrated, interprofessional strategy and directly referenced the contributions of complementary and integrative therapies and practitioners.

Most recently, with support from the Lotte and John Hecht Memorial Foundation, ACCAHC became a sponsoring organization of the 2012-2014 Institute of Medicine Global Forum on Innovation in Health Professional Education. Dr. Goldblatt is the IOM forum member, with John Weeks also active as an alternate. Among positive outcomes from the early engagement is a general understanding that optimal team care will involve "widening the circle" of professionals and stakeholders who are considered part of these teams. ACCAHC's Board charged the Goldblatt-Weeks team to work on elevating consideration of innovations that lead to education of healthcare professionals who can not only react to and manage disease but also partner with patients and communities to promote practices and choices that will lead to health and wellness. Meantime, ACCAHC is developing efficient means of translating the leading-edge thinking and content to its councils of schools and their individual school and program members.

More information on ACCAHC's areas of work, and accomplishments to date, are listed in Appendix 2, ACCAHC Accomplishments: January 2005-January 2013. A narrative of ACCAHC work in collaborative activity from 2005-2009 is found in Appendix 3, entitled: Toward Collaboration between the Disciplines. Up-to-date information on ACCAHC's directions and key personnel are on the ACCAHC website at www.accahc.org. We welcome your queries made through info@accahc.org.

Appendix 2

ACCAHC Accomplishments: January 2005-January 2013

Providing a Shared Voice for Integrative Health and Medicine

IOM's 2009 National Summit on Integrative Medicine and Health of the Public	ACCAHC secured an invitation from the Institute of Medicine which ensured that ACCAHC member organizations would be represented on the 14 member Planning Committee for the February 25-27, 2009 Summit which was sponsored through a grant from the Bravewell Collaborative. The ACCAHC nominee who was appointed, Elizabeth Goldblatt, PhD, MPA/HA, was the only representative from non-conventional healthcare on the Committee. The meeting was significantly more multi-disciplinary in tone in part through Goldblatt's representation of patient interest in the full scope of practitioners involved in integrative medicine.
Two National Pain Initiatives: IOM and Center for Practical Bioethics	ACCAHC secured an invitation from the Institute of Medicine which ensured that ACCAHC member organizations would be represented on the IOM Committee on Pain Research, Care and Education. Pain expert Rick Marinelli, ND, LAc was selected to serve on the Committee. ACCAHC was invited to represent patient interest in complementary health care as a member of the Pain Action Alliance to Implement a National Strategy (PAINS) led by the Center for Practical Bioethics. Martha Menard, PhD, CMT sits on the PAINS steering committee and chairs a multidisciplinary ACCAHC Task Force on Integrative Pain Care that is informing her work. The Task Force is developing a paper on the potential value of complementary and integrative approaches in more appropriately using opioids.
Participating in the development of the NIH NCCAM 2011-2015 Strategic Plan: Setting Research Priorities	The ACCAHC Research Working Group (RWG) led ACCAHC into continuing dialogue with the NIH National Center for Complementary and Alternative Medicine. The ACCAHC focus is on the kind of "real world research" that is likely to most inform the ability of consumers and other stakeholders considering services from these disciplines. The RWG was particularly involved in the NIH NCCAM 2011-2015 Strategic Plan. In a series of letters, phone conferences and a face to face meeting, ACCAHC has underscored the importance of looking at whole practices, at cost issues and in building the capacity of these disciplines to participate in the research endeavor. See: http://accahc.org/images/stories/2011-15_strat_plan.pdf. A key role was in elevating the importance, as underscored by Congress, of looking not merely at the impact of individual therapies, but at "disciplines" from which consumers typically receive treatment. Work was led by Greg Cramer, DC, PhD and Carlo Calabrese, ND, MPH. The final NCCAM plan substantially reflected ACCAHC recommendations in this area.

Patient-Centered Outcomes Research Institute (PCORI)	Through leadership of the Research Working Group (RWG), ACCAHC provided public comment on the new Patient-Centered Outcomes Research Institute (PCORI) established under the Affordable Care Act. In addition, two ACCAHC leaders were part of a panel asked to present on complementary and integrative medicine at a September 2011 Listening Session for the PCORI Board of Governors. RWG founding co-chair and member *emeritus*, Christine Goertz, DC, PhD, was appointed to serve on the PCORI Board. The RWG is also supporting submission of a proposal on practice-based research network for integrative practices, and engaged in a dialogue with PCORI CEO Joseph Selby, MD, MPH, the recording of which is here: http://accahc.org/images/Selby.mp3.
The Movement for Inter-professional Practice and Education	ACCAHC early recognized the importance of the value national and international movement toward team care and interprofessional education and practice (IPE/P) both for enhancing care and as an opening to include the patient-centered interest in health-focused treatment and complementary and integrative practices and disciplines. ACCAHC has: participated in the Collaboration Across Borders conferences (2011, 2013); development relationships with the Interprofessional Education Collaborative formed by academic councils for medical doctors, nurses, public health, osteopathy, dentistry and pharmacy; developed relationships with leaders of the National Center for Interprofessional Practice and Education and sponsored an IOM Global Forum related to the subject. (See below.) Our view is simple: in patient-centered movements such as this, the interest of patients in integrative health practices and disciplines must be at the table.
Institute of Medicine (IOM) Global Forum on Innovation in Health Professional Education	In early 2012, the ACCAHC Board of Directors chose to invest in ACCAHC becoming a founding co-sponsor of the 3-year Institute of Medicine Global Forum on Innovation in Health Professional Education. ACCAHC is the sole integrative health-focused sponsor of the forum, joining over 30 councils of colleges and professional associations from medicine, nursing, public health, pharmacy and other disciplines. ACCAHC's priority is to collaborate with the other forum members to explore innovations associated with graduating students who are expert in assisting individuals and communities toward wellness, self-care, personal empowerment, prevention and health promotion. Participation is made possible in part via an anonymous foundation grant. An IOM "Spotlight" on ACCAHC's Forum member Elizabeth Goldblatt, PhD, MPA/HA is here: http://iom.edu/Activities/Global/InnovationHealthProfEducation/Member-Spotlights/Goldblatt.aspx.
Expanding External Affairs 'Ambassadors" Program	Recognizing both the opening opportunities for participation of individuals representing ACCAHC's values and disciplines in numerous healthcare initiatives, ACCAHC's Board formally endorsed pro-active outreach to place individuals in key dialogues, projects and initiatives. Among those with formal connections as of January 2013: Integrative Medicine for the Underserved; International Congress for Clinicians in Complementary and Integrative Medicine; Pain Action Alliance to Implement a National Strategy; and the HRSA-funded National Coordinating Center for Integrative Medicine of the American College of Preventive Medicine.

Publication of a Critically Needed Educational Tool	ACCAHC created the Clinician's and Educator's Desk Reference on the Complementary and Alternative Healthcare Professions to support all practitioners in understanding the value of the whole practices – not just therapies – our disciplines represent. Content is developed by each council of colleges. The book has been widely heralded across multiple disciplines. See quotes about the book here: http://accahc.org/images/stories/cedr_quotes_050511.pdf.
Integrative Health and Medicine in Accountable Care Organizations (ACOs) and Patient-Centered Medicine Homes (PCMHs)	The movement in policy, payment and practice toward models of care that allow payment for helping people to be healthy, avoiding costly procedures and surgeries, and stay well created an opening for complementary and integrative practices and providers. ACCAHC has developed a significant initiative in this area, related to Competency #6 (see below) of ACCAHC's competencies, via a task force that includes a former Centers for Medicare and Medicaid (CMS) senior policy member who worked on the ACO language and an integrative medical doctor with significant experience as a health system quality leader. The project is *Strategic Support for Stakeholders in Exploring Integrative Health and Medicine in Affordable Care Organizations and Patient-Centered Medical Homes.*

Engaging Relationships with MD-RN-PhD Educators in Integrative Health

National Education Dialogue (NED)	Led various projects for the National Education Dialogue to Advance Integrated Health Care: *Creating Common Ground* (2004-2006) which involved educators from 12 disciplines and focused on leaders of Consortium of Academic Health Centers for Integrative Medicine (CAHCIM).
More Inclusive Definition of Integrative Medicine	Collaborated with the now 54-member CAHCIM to change their definition of "integrative medicine" to better respect the integration of whole disciplines by including the importance of integrating not just modalities but also healthcare professionals and disciplines.
Research Conferences	ACCAHC was a Participating Organization for the May 2009 North American Research Conference on Complementary and Integrative Medicine (NARCCIM), as it was for the 2006 meeting and again for the 2012, renamed, International Research Congress on Integrative Medicine and Health. ACCAHC leaders have served on key planning committees, prepared and had accepted multiple programs for presentation, and reported ACCAHC survey research at poster sessions.
Published Response on "Competencies in Integrative Medicine"	Publication of a peer-reviewed paper which underscored the importance of integrating disciplines and practitioners, not just therapies, in "integrative medicine."(Response to a Proposal for an Integrative Medicine Curriculum, Journal of Alternative and Complementary Medicine. November 1, 2007, 13(9): 1021-1034)

Ongoing Dialogue with Integrative Medicine Educators	ACCAHC has fostered and maintained collegial relationships with CAHCIM. A highlight was a May 2009 gathering of roughly 40 people from each organization in shared working groups after which CAHCIM sponsored a reception and dinner. Key CAHCIM leaders are on ACCAHC Council of Advisers and informed dialogues at the 2011 ACCAHC Biennial Meeting. ACCAHC working group members are working on CAHCIM-originated projects and CAHCIM participants are working on some that originated with ACCAHC.
International Congress for Educators in Complementary and Integrative Medicine	ACCAHC co-sponsored with CAHCIM and Georgetown University for the first International Congress for Educators in Complementary and Integrative Medicine October 24-26, 2012. This exemplar of "horizontal integration" included interprofessional teams on the planning committee and program committee and content distributed across various disciplines. ACCAHC's Education Working Group co-chair Mike Wiles, DC, MEd co-chaired the program committee and ACCAHC EWG co-chair Jan Schwartz, MA, was among those who served on it. Over 340 educators, clinicians and researchers attended. ACCAHC staff directly supported seven different presentations on such topics as competencies for practice in integrative environments.
National Coordinating Council for Integrative Medicine/American College of Preventive Medicine	ACCAHC supported a successful 2012 grant proposal of the American College of Preventive Medicine (ACPM) that led to an award from the US Health Resources and Services Administration (HRSA) to create a National Coordinating Center for Integrative Medicine. ACCAHC's current and founding executive directors, John Weeks and Pamela Snider, ND, respectively, serve on the steering committee for the project, as does Council of Advisers member Len Wisneski, MD.

Strengthening Collaboration among ACCAHC's Disciplines

Interprofessional Collaboration Among the Whole Person-Oriented Disciplines	While lumped as "CAM," the disciplines that share in common a whole person and health-focused view of health care have not historically collaborated closely to advance their shared values and projects. Via scores of conference calls and numerous face-to-face gatherings (February 2005, June 2005, May 2006, October 2007, July 2008, May 2009, May 2010, June 2011, October 2012), now 17 key national and North American organizations across our disciplines have committed and re-committed year-after-year to collaboration as dues-paying members of a shared organization ($1000-$5000/year, depending on size).
Organized Special Interest Groups (SPIGs)	In order to foster mutual understanding and to enhance the abilities of our distinct organizations, ACCAHC convenes via phone conferences and face-to-face meetings SPIGs for Councils of Colleges, for Accrediting Agencies, for Certification and Testing Organizations, and Traditional World Medicines/Emerging Professions.

Presentations and Publications	ACCAHC survey projects have been the subject of various posters and presentations: the 2006, 2009 and 2012 scientific meetings of the North American Research Conference on Complementary and Integrative Medicine (NARCCIM); the 2007 meeting of the Society for Acupuncture Research; the *Journal of Alternative and Complementary Medicine*; and the 2012 International Congress for Educators in Complementary and Integrative Medicine.
Researching Competencies for Integration	In a grant-funded project from the National Certification Commission for Acupuncture and Oriental Medicine, ACCAHC, working with others, carried out two surveys on competencies of licensed acupuncturists for practice in MD-dominated settings. A project on Hospital Based Massage Therapy is engaging a survey (2013) to support development of competencies for massage practice in hospital settings.
Clarified Priority Projects	Once convened, ACCAHC identified core, major, multi-year projects in which members and participants could collaborate in advancing patient care through advancing mutual respect. Key areas identified as of January 2011 are: 1) supporting students, educators, clinicians and other stakeholders in working optimally in integrated environments; 2) supporting educators in developing evidence-informed education (via a collaboration with schools that received NIH-NCCAM R-25 grants to enhanced evidence-based programs); 3) engaging the national dialogue to foster interprofessional education/care; and 4) developing programs and resources to support leadership in integrative health care.
Competencies for Optimal Practice in Integrated Environment	To support priority project #1, above, ACCAHC engaged an 11-month process involving over 50 professionals from 8 disciplines to create a document entitled Competencies for Optimal Practice in Integrated Environments. Work was led by working group co-chairs Jan Schwartz, MA, Mike Wiles, DC, MEd, MS, Marcia Prenguber, ND and Jason Wright, LAc . ACCAHC leaders then merged this document with the subsequently-published Core Competencies for Interprofessional Collaborative Practice of the Interprofessional Education Collaborative (IPEC). These Competencies, now endorsed by most of the ACCAHC councils of colleges, have been reported in various meetings and shared with the IPEC leadership. Work is underway to create materials and programs, such as the ACO/PCMH project above, to translate the Competencies into practice.
Cross-Disciplinary Working Groups	ACCAHC's core structure is based in 3 multidisciplinary committees of leaders across our disciplines: Education Working Group, Clinical Care Working Group and Research Working Group. In each, interprofessional teams independently set priorities that are aligned with the ACCAHC vision and mission. A multitude of projects have been engaged.

Connecting Educators and Large Employers	Members of the ACCAHC Research Working Group presented a special workshop to an April 2008 Institute for Health and Productivity Management conference of large employers on the potential value of integrative services in the health of employees. The IOM Global Forum participation heightens the importance of all health professional education connecting more closely to real world interests. The ACO/PCMH project speaks directly to payment and delivery models long supported by employers.
Visibility of the Mission	ACCAHC was the subject of a May 2008 article in the peer-reviewed *Explore: The Journal of Science and Healing!* ACCAHC's work has since been referenced in multiple media including *Acupuncture Today, Integrative Medicine: A Clinician's Journal, The Huffington Post, Massage Today, Dynamic Chiropractic, International Association of Yoga Therapists*, and others.
Philanthropic Support: *Breaking the Glass Ceiling*	ACCAHC has begun developing a philanthropic base. Sustaining Donors have made 3-year pledges totaling up to $50,000/year. ACCAHC's major initiative, the Center for Optimal Integration, will require that it break the "glass ceiling" that has kept significant philanthropy from investing in integrative projects that are not led by either medical doctors or, in rare cases, by nurses. In late 2011, the Westreich Foundation granted ACCAHC $100,000 over 3 years, the first major commitment following ACCAHC's support from its founding philanthropic partner, Lucy S. Gonda. ACCAHC is actively seeking additional philanthropic leaders who will help remove this barrier to creation of optimal care.

Appendix 3

Toward Collaboration between the Disciplines: An ACCAHC Perspective

This chapter was adapted and updated from content ACCAHC made available for a textbook entitled *Collaboration across the Disciplines in Health Care*[1], which ran as a chapter titled "Complementary, Alternative and Integrative Health Care Perspectives." The chapter was co-authored by members of the ACCAHC executive committee at that time: John Weeks, Elizabeth Goldblatt, PhD, MPA/HA, Reed Phillips, DC, PhD, Pamela Snider, ND, Jan Schwartz, CMT, David O'Bryon, JD, Marcia Prenguber, ND, and Donna Feeley, MPH, RN, LMT.

Overview

This Desk Reference is offered as an educational tool about the licensed complementary healthcare professions and related disciplines. Yet our purpose is more fundamental. Our goal is to create a basis for better understanding and mutual respect between members of distinct healthcare disciplines, so that the health care patients receive will be enhanced. Collaboration is the primary tool through which we will achieve that goal.

This chapter explores collaborative activity between educators in conventional academic medicine and educators from the disciplines represented in this text, particularly the five disciplines with federally-recognized accrediting agencies. These are acupuncture and Oriental medicine (AOM), chiropractic, naturopathic medicine, direct-entry (home-birth) midwifery, and massage therapy.

For these disciplines, significant local and national dialogue with conventional medical educators, and among themselves, emerged formally with the turn of the 21st century. We begin with a description of some of the historical context for this activity, including recommendations from the White House Commission on Complementary and Alternative Health Care Policy[2], the Institute of Medicine[3], and the multi-stakeholder and multidisciplinary National Policy Dialogue of the Integrated Healthcare Policy Consortium.[4]

We then focus on the actions and directions set by the collaborative National Education Dialogue (NED) to Advance Integrated Health Care: *Creating Common Ground*[5] and the Academic Consortium for Complementary and Alternative Health Care (ACCAHC). We also note some cross-disciplinary educational initiatives of some members of the Consortium of Academic Health Centers for Integrative Medicine (CAHCIM), a group of now 54 conventional medical institutions active in the integrative medicine field. This chapter includes several sidebars that shed useful light on the activities of some of these entities, and identifies priorities for action collaboratively established by these mixed groups of academics.

Finally, we have included some clarifying values, recommendations, and challenges encountered on this mission to improve health care.

Dynamics Shaping the Emerging Interest in Collaboration

As noted in the Introduction, the period from 1980 to the present has been one of significant expansion for schools, accreditation, licensing, and self-regulatory efforts among the licensed complementary healthcare professions. Communication and collaboration between the complementary and alternative healthcare disciplines and conventional medical professionals have expanded over these decades, yet effective, widespread engagement has only commenced in the last 15 years. A long history of exclusion and antagonism separated biomedicine from disciplines promoting holistic, integrative and natural healthcare practices. Attitudes developed during that period still influence and sometimes dominate the dialogue. Residual misunderstandings, mistrust, and often ignorance originating during that period of extreme polarization can still shadow efforts to create greater collaboration. But much new openness presently exists.

Table A.2 offers a view of how that history of exclusion created attitudes and perceptions that can still affect interdisciplinary relationships. The bottom line of the table, as with past interdisciplinary experiences, is that both conventional and complementary or integrative health providers typically saw patients who were the failures of their counterparts on the other side of the divide. Images of the other disciplines were, therefore, formed accordingly.

Fortunately, the rift is on the mend. Many factors, including general sociocultural shifts toward holism, environmentalism, internationalism and multicultural awareness, created fertile ground for collaboration to take root and flourish. How this new acceptance originated continues to shape the nature of current collaborative activity. The following developments, therefore, merit special recognition.

Opening through force: An antitrust lawsuit
In 1990, the chiropractic profession concluded a 15-year antitrust battle against the American Medical Association (AMA), *Wilk vs. the AMA*. The chiropractors' grassroots campaign led to a determination that the AMA had engaged in illegal restraint of trade through such practices as excluding chiropractors from hospitals and forbidding AMA members from sharing patients with chiropractors. Ending the era in which the AMA branded not just chiropractors but all non-conventional practitioners as quacks or frauds created a potential for better-informed dialogue between conventional medicine and all of the alternative and holistic fields.

Understandably, the attitudes and feelings engendered by discriminatory actions are still present today. Medical doctors educated in the earlier era are now in leadership positions in academic health centers and hospitals. Similarly, many leaders in the complementary and alternative healthcare professions entered their training in a highly adversarial atmosphere. While both conventional and integrative health professionals from that earlier time are embracing the integration process, younger practitioners are typically more influenced by recent cultural trends and are more likely to have personally used integrative approaches in their own care or for that of family members.

Public interest and demand

An extremely powerful force generating early momentum toward improving the flow of information and communication was the evidence that over a third of consumers in 1991 used some form of unconventional medicine and most chose not to tell their conventional providers about it[6]. Originally reported in the *New England Journal of Medicine* in January 1993, then updated for the *Journal of the American Medical Association* in 1998[7], the survey findings stimulated a fundamental shift in perception among public officials, politicians, and members of the media. Alternative medicine users, it turned out, were everywhere.

Hospitals, academics, insurers, politicians, and employers began exploring how to take advantage of this pervasive consumer interest. Subsequent national surveys have suggested that between 40% and 85% of healthcare consumers with chronic conditions typically try some form of non-conventional therapies or disciplines. The level of consumer use and estimates of cash payments for these services piqued the interest of conventional stakeholders.

Table A.2

Bilateral Prejudice as an Operational Issue Limiting the Integration of Complementary and Conventional Health Care

Phenomenon	Conventional Perspective	Complementary Perspective
Successful complementary treatment	placebo for self-limiting condition	proof of value of complementary care
Successful conventional treatment	proof of value of conventional care	suppresses problems / fails to address causes
Science in conventional medicine	our foundation	80-90% of procedures have no support for clinical efficacy
Science in complementary medicine	virtually non-existent	strong in some areas / growing body of evidence
Conventional primary care formulary	good tools	70% of problems go away on their own
Complementary primary care formulary	70% of problems go away on their own	good tools
Complementary diagnostics	of great concern / likely to misdiagnose	consider more than lab values
Conventional therapeutics	best in the world	often inappropriately invasive / likely to miss causes
Surgeon quick to operate	overutilizer	rip off artist
Chiropractor always requesting 20 visits	rip off artist	overutilizer
Antibiotics for viral conditions	basically harmless treatment demanded by patients	harmful to the individual and the population
Complementary therapies in end-stage terminal conditions	waste of money/creates false hope	alleviate side-effects of conventional treatment/help patients with transition
Oncologist creating new chemotherapeutic "cocktails"	brilliant clinician	experimenting with poison
Naturopath who mixes Asian and homeopathic treatments	totally wacko	brilliant clinician
Self-healing	self-limiting condition, spontaneous remission	healing power of nature
Clients of the "other" who come to my office	their failures	their failures

Note: The table was created by John Weeks in 1996 from data acquired from surveys and interviews. While it does reflect early perceptions that framed the integration dialogue, the choice of language does not reflect ACCAHC's current values and approaches. In addition, the phrases used in the table did not reflect the judgments of all providers from either group; they were generalizations.

Source: Operational Issues in Incorporating Complementary and Alternative Therapies and Providers in Benefit Plans and Managed Care Organizations. Prepared for the Workshop, Complementary and Alternative Medicine: Issues Impacting Coverage Decisions, sponsored by the US National Institutes of Health, Office of Alternative Medicine, US Agency for Health Care Policy and Research, and the Arizona Prevention Center at the University of Arizona Health Sciences Center; October 9, 1996; page 45. © John Weeks. Reprinted with permission.

Introduction of significant research and education grants
In 1991, via a $2 million Congressional earmark sponsored by Senator Tom Harkin, the NIH was urged to begin exploring what was then termed unconventional medicine. This initiative became the NIH Office of Alternative Medicine, which became the National Center for Complementary and Alternative Medicine (NCCAM) in 1998. The nation's academic health centers, which had heretofore shown little interest in these fields, saw that federal money was a significant new factor in what had originated primarily as a consumer-led movement. New grant funding also supported the interests of conventional academics who saw value in researching nonconventional practices.

NCCAM's annual budget was roughly $120 million in 2008, most of which was distributed as grants to conventional medical centers. Over a dozen significant grants have gone to conventional academic health centers for education-related projects, most to members of the Consortium of Academic Health Centers for Integrative Medicine. Some of these conventional medical schools have had licensed complementary healthcare educator-researchers as advisers; some had AOM, chiropractic, massage or naturopathic institutions as partners. In many of these grants, NCCAM required applicants to have partnerships with complementary and alternative healthcare institutions. Likewise, NCCAM subsequently awarded a series of grants to complementary healthcare academic institutions, with a requirement that they have conventional partners. Most of the current best practices in inter-institutional, collaborative action were initiated through these NCCAM grants.

An ACCAHC analysis in 2010 estimated that just 4.6% or $60-million of total NCCAM funding between 1999 and 2010 went directly to institutions that educate the distinctly licensed integrative health disciplines represented in ACCAHC's membership. ACCAHC has sought to educate the NIH to the importance of increased direct granting to these institutions as a critical foundation for building the scientific infrastructure that is needed to underwrite optimal integration.

Despite the limits in direct granting, a major influence for stimulating collaboration was the infusion of federal money from NCCAM, which had been foisted on the NIH by Congressional action.

Emergence of integrative medicine in conventional academic health centers

Closely associated with the availability of research funding was the development of the concept of, and expansion of interest in, what is now called integrative medicine. In 1991, just one conventional medical school was known to have any education relative to complementary and alternative medicine. By 1998, that number had reached 75. The Rosenthal Center for Complementary and Alternative Medicine compiled an extensive list of conventional schools offering course work in CAM, and made the following comment on its website: "At the time we archived the resource (June 2007) it is rare that any conventional medical school would not have courses in complementary, alternative, or integrative medicine."

In 2001, the Consortium of Academic Health Centers for Integrative Medicine (CAHCIM) was formed by nine academic health centers. To join, a center needed to have institutional support at the dean's level or above, and integrative medicine activity in at least two of the three core areas of healthcare academics (education, research, clinical services). By 2013, the number of CAHCIM members throughout North America had grown to 54. An organization of philanthropists, the Bravewell Collaborative, provided significant backing for the emerging consortium. The development of CAHCIM created a new focal point for collaborative academic activity and shifted the landscape in alternative medicine integration.

Each of these factors opened doors for change agents, inside and outside of mainstream medicine, who believed that health care could be improved through fostering collaboration between conventional professionals and members of disciplines that had historically been ostracized. Most of these professionals were integrative-oriented medical doctors, nurses, or administrators, some of them cross-trained in a complementary and alternative healthcare discipline or modality. Evidence of consumer use allowed these individuals to more publicly promote dialogue, integration, and collaboration.

Yet the ongoing work of these collaboratively-minded professionals, and their partners in the complementary and alternative healthcare disciplines, continues to be challenged by weak institutional support and, in some cases, philosophical differences. The CAHCIM definition of integrative medicine is clearly interprofessional, calling for the integration of all "appropriate healthcare approaches, health professionals and disciplines," while other integrative medicine definitions focus on only integrating new therapies into conventional practices. For instance, an integrative medicine specialty being developed through the American Board of Physician Specialties does not reference the importance of working in teams as a core principle.

Optimally, integrative action would be advanced by a desire to understand why patients are drawn to these services, and how care outcomes could be bettered, and/or medical costs reduced, by understanding and integrating these heretofore excluded practitioners. However, the ongoing institutional commitments needed to support significant educational, economic, and cross-cultural integration initiatives has often been minimal, making progress very slow.

Promptings from a White House Commission, the Institute of Medicine, and a Policy Initiative

Beginning in 2001, three significant national initiatives provided intellectual and policy-related support for those who chose to begin promoting more exploration of interdisciplinary activity between conventional and complementary and alternative healthcare professions. The report of the White House Commission on Complementary and Alternative Medicine Policy (NIH, 2002) urged that:

> ... agencies should convene conferences of the leaders of CAM, conventional health, public health, evolving health professions, and the public; of educational institutions; and of appropriate organizations to facilitate the establishment of CAM education and training guidelines. Subsequently, the guidelines should be made available to the states and professions for their consideration.[2(pg. 99)]

The Commission endorsed ten guiding principles. One recognizes that "partnerships are essential to integrated health care."[2 (pg. xxiii)] Cross-disciplinary understanding and respect are underscored: "Good health care requires teamwork among patients, healthcare practitioners (conventional and CAM) and researchers committed to creating optimal healing environments and to respecting the diversity of all health care traditions." [2(pg. xvi)]

The 2001 National Policy Dialogue to Advance Integrated Care: *Finding Common Ground*, which also published its findings in 2002, made similar recommendations. This multidisciplinary and multi-stakeholder gathering of 60 leaders in integrated care reached consensus on a core recommendation to "Establish a national consortium of conventional and CAM educators and practitioners." Such a consortium would:

... identify a core education for all conventional and CAM educational institutions, created by educators from the respective institutions themselves to ensure that the healthcare professionals of the future understand what all systems of health care can offer. This will result in healthcare providers, both conventional and CAM, learning at an early point in their training what other systems and modalities offer. The consortium will develop educational standards to ensure that factual and accurate information is taught across all disciplines and that the type of information is in accordance with the teachings and principles of the respective disciplines.[4(pg. 12)]

The policy-oriented gathering identified a possible need for a special accrediting entity for continuing education and ongoing governmental support for collaborative education:

An interdisciplinary body will be created to accredit continuing education programs in CAM/IHC for all medical disciplines so that medical, nursing, pharmacy students and other health professions students and practitioners, whether conventional or CAM, continue to learn about and respect the value of CAM and integrative healthcare approaches. Federal support is necessary to fund training opportunities in integrative healthcare settings so that medical students and healthcare practitioners, whether conventional or CAM, can learn about integrative approaches in order to provide better and more cost-effective patient care.[4(pg. 11-12)]

Finally, in 2005, the Institute of Medicine published a *Report on Complementary and Alternative Medicine in the United States*, which acknowledged both the growth of the field and the central importance of collaborative and interdisciplinary action:

The level of integration of conventional and CAM therapies is growing. That growth generates the need for tools or frameworks to make decisions about which therapies should be provided or recommended, about which CAM providers to whom conventional medical providers might refer patients, and the organizational structure to be used for the delivery of integrated care. The committee believes that the overarching rubric that should be used to guide the development of these tools should be the goal of providing comprehensive care that is safe and effective, that is collaborative and interdisciplinary, and that respects and joins effective interventions from all sources.[3(pg. 216)]

The report speaks directly to the problems that may arise if integrative medicine confines itself to the best uses of distinct approaches and therapies instead of honoring a medical pluralism that respects significant differences in "epistemological framework(s)" of practitioners from different disciplines.

Yet, care must be taken not to assume compatibility where none exists. Medical pluralism should be distinguished from the cooptation of CAM therapies by conventional medical practices. (Kaptchuk and Eisenberg, 2001)

Although some conventional medical practices may seek and achieve a genuine integration with various CAM therapies, the hazard of integration is that certain CAM therapies may be delivered within the context of a conventional medical practice in ways that dissociate CAM modalities from the epistemological framework that guides the tailoring of the CAM practice. If this occurs, the healing process is likely to be less effective or even ineffective, undermining both the CAM therapy and the conventional biomedical practice. This is especially the case when the impact or change intended by the CAM therapy relies on a notion of efficacy that is not readily measurable by current scientific means.[3(pg. 175)]

It is noteworthy that, while this comment urges respect for distinctions between the disciplines, the report itself focused significantly more on integration of therapies than on integration of disciplines. A search by medical journalist Elaine Zablocki found that, in the IOM report, "CAM therapies" appears 299 times, "CAM practitioners" 66 times, "CAM professions" is used eight times, and "CAM disciplines" is used twice. The internal dialogue that was framed was one of grafting truncated therapies rather than the integration of disciplines, with all of their distinct differences and cultures.

While the decisions of many consumers of medicine have already provided ample reason for members of our diverse disciplines to learn to interact more closely, these three national initiatives added necessary perspectives on policy and medical leadership.

Initial Conventional Explorations in Collaborative Education

Between 2000 and 2005, 14 conventional medical and nursing schools and one family medicine residency program received multi-year grants from the NIH NCCAM as part of the Complementary and Alternative Education Project. The October 2007 issue of *Academic Medicine* (Vol. 82, No. 10) devoted 50 pages to nine reports based on these experiences. Of particular relevance to ACCAHC and educators in complementary and alternative healthcare institutions is an article by Anne Nedrow,

MD and others about the four academic health centers that formed collaborative partnerships with nearby academic institutions that train CAM practitioners[8].

Developers of these collaborative programs note that respect for the potential value of complementary and alternative healthcare practitioners is enhanced by giving conventional practitioners direct experience of these approaches in educational, clinical, and research settings. Medical students at Georgetown University, for instance, report surprise in learning, during shared gross anatomy lab, the high level of understanding of anatomy that massage therapists are required to achieve. Confidence of medical students in working with complementary healthcare practitioners increased significantly with personal experience of the therapies among a group of medical students at the University of Minnesota[9].

Nedrow's integrative medicine program at the Oregon Health and Science University forged ties with Western States Chiropractic College (now called the University of Western States), National College of Natural Medicine (where the dominant program is naturopathic medicine), and Oregon College of Oriental Medicine to form the Oregon Collaborative for Integrative Medicine. Nedrow's conscious effort to put all of the participants in a non-hierarchical relationship, dubbed "lateral integration," was well-received by the complementary healthcare educators. Nedrow has reported that barriers between professionals of different disciplines can be diminished when it is educators who are seeking to collaborate. The shared mission of educators to optimize the education of their students becomes the basis for open-minded collaboration.

Nedrow and her co-authors anticipate that such collaborations will increase. A key emerging issue they identified is credentialing of complementary healthcare practitioners to teach within conventional health professions institutions. The authors recognize that collaboration involves development of respect even in a context of significant differences. One proposed solution is to teach complementary health care "as an aspect of cultural sensitivity."

One sign of emerging openness to collaboration with members of other disciplines is that, by 2007, schools belonging to CAHCIM (for which Nedrow and her co-authors have been leaders) included faculty members who are also licensed members of the chiropractic, acupuncture and Oriental medicine (AOM), and naturopathic medical professions. These dually credentialed educators are presently serving on CAHCIM working groups and participating in their annual conferences. Beginning in 2009, three of these licensed complementary healthcare professionals served CAHCIM as co-chairs of their working groups. In 2013, a naturopathic physician, Ather Ali, ND, MPH, represented Yale Integrative Medicine on the CAHCIM steering committee.

Engaging a National, Multidisciplinary Collaboration

In 2004, amidst this rapidly developing environment, the National Education Dialogue to Advance Integrated Health Care: *Creating Common Ground* (NED) was engaged by the Integrated Healthcare Policy Consortium, which birthed ACCAHC. By design, many of the leaders of CAHCIM, as well as the founders of ACCAHC, were involved in the 25-member planning team. NED ultimately included educators from 13 disciplines at a four-day meeting of 75 educators in mid-2005 (see Table A.3). A parallel process involved the creation of ACCAHC during the same time period (described in Appendix 1).

Table A.3

Disciplines of Educator Participants in the National Education Dialogue

Acupuncture and Oriental medicine*	Chiropractic*
Holistic medicine	Holistic nursing
Homeopathic medicine	Integrative medicine
Massage therapy*	Midwifery (direct-entry)*
Naturopathic medicine*	Nutrition/dietetics
Occupational medicine	Osteopathic medicine
Public health	Yoga therapy**

*ACCAHC core disciplines.
**Yoga therapy, while not a licensed practice, is involved in ACCAHC as an emerging, Traditional World Medicine discipline.

The 15 months of NED planning revolved around identification and development of a series of projects designed to provide content for the onsite meeting to consider. One project was a survey of all of CAHCIM's integrative medicine programs (then 28) and all of the accredited schools and programs from the ACCAHC disciplines (then 130) to gain baseline data on the "Status of Inter-Institutional Relationships between CAM Disciplines and Conventional Integrative Medicine Programs." Sample findings:

- Few of the then-28 CAHCIM member programs had formal classroom or clinical relationships with any complementary and alternative healthcare schools or colleges. The greatest involvement was found with AOM schools (32% for classroom, 16% clinical) and massage schools (20% classroom, 20% clinical).
- Belief that developing "stronger, multi-dimensional, interdisciplinary relationships" is key to creating a "fully integrated healthcare system" was strong; 85% of CAHCIM respondents and 86% of the ACCAHC respondents agreed.
- Sharing documents and strategies on best practices was viewed as the most beneficial strategy to support more inter-

institutional relationships (73% of CAHCIM and 77% of ACCAHC).

In addition, a set of survey questions directed only at ACCAHC programs found that a majority of respondents had diverse clinical care relationships with conventional delivery organizations in their communities. Among these were community clinics (48%), hospitals (28%), senior homes (43%), and homeless clinics (33%). Such participation was highest in the naturopathic medical field, followed by chiropractic and AOM. The data suggested that integrated clinical services might take place in these off-campus environments, many of which were already linked to conventional residencies and clinical training experiences.

Clarifying the core competencies of integrative medicine

A second, multidisciplinary survey was stimulated by the response of ACCAHC participants to the publication of a CAHCIM-endorsed paper in *Academic Medicine* on competencies in integrative medicine.[10] The paper was developed internally by CAHCIM, without input from leaders of the ACCAHC disciplines. A modified Delphi survey administered to a subset of 18 complementary healthcare educators representing all the ACCAHC core disciplines eventually elicited five key areas of concern with the competencies endorsed by CAHCIM.[11] Most of the concerns relevant to the core issue of integration were not focused solely on therapies, but on an active engagement between diverse disciplines. The survey found that:

- The CAHCIM paper leaves the impression that conventional medical physicians may simply incorporate into their practices what they perceive to be good CAM therapies rather than referring to or co-managing and collaborating with CAM providers.

- The CAHCIM paper does not include the option of integrated care with MDs and CAM practitioners as partners, and seems to propose that CAM be an add-on to conventional medical care.

- Occasionally in the paper, language appears to reflect a continuation of the 'us and them' mindset rather than seeing that CAM providers, faculty, and systems could and should be part of IM.

- The paper gives the impression that conventional medical institutions want to include CAM, but not necessarily CAM practitioners, in their vision of IM.

- CAM is defined in relationship to biomedicine as complementary or alternative, but is considered integrative if delivered by conventional physicians.[11(pg. 1023)]

The issue of the distinction between integration of therapies versus the more challenging and the significantly more beneficial work of collaboration between disciplines led to a process in which CAHCIM changed its definition of integrative medicine to better reflect the importance of working with other disciplines (see Sidebar A.1).

Efforts to create shared values to guide interdisciplinary action

A third project, led by holistic nurse educator Carla Mariano, EdD, RN, AHN-BC, FAAIM, sought to clarify a shared set of values across the disciplines involved with the NED process. The vision behind this project was of a document that might be co-produced by a multidisciplinary NED team and endorsed by the NED planning committee and then amended and endorsed by the onsite gathering of 75 educators. If agreement were reached, the document could then be taken to the individual disciplines for endorsement as a shared basis for interdisciplinary action. The operating belief of the ten educators from eight disciplines who comprised the Values, Knowledge, Skills and Attitudes

---
Sidebar A.1

From Integration of Therapies to Collaboration among Disciplines

The planning team for the National Education Dialogue (NED) to Advance Integrated Health Care: *Creating Common Ground* was an alphabet soup of titles never before gathered for a shared educational purpose, including MD, ND (naturopathic physician), LAc (acupuncture and Oriental medicine practitioner), MPH, DC (doctor of chiropractic), PhD, LM (licensed midwife), RN, LMT (licensed massage therapist), and RYT (registered yoga teacher).

The hope and expectation among educators and practitioners was that such educational collaboration would translate into more understanding and greater cooperation among practitioners. The organizing mantra was *those who learn together practice together*. What could educators do to help get their disciplines out of separate silos and break down the historic barriers between conventional and alternative professions? An early step was to clarify the shared vision, if any, unifying the leaders. The group agreed on this statement:

We envision a healthcare system that is multidisciplinary and enhances competence, mutual respect and collaboration across all complementary and alternative medicine and conventional healthcare disciplines. This system will deliver effective care that is patient centered, focused on health creation and healing, and readily accessible to all populations.[5](pg. 1)

A funny thing happened on the way to working on this vision. On the December 2004 NED planning team call, one of these planners announced that at the recent general meeting of CAHCIM, the conventionally-based academics had agreed on a definition of integrative medicine.

Integrative medicine is the practice of medicine that reaffirms the importance of the relationship between practitioner and patient, focuses on the whole person, is informed by evidence, and makes use of all appropriate therapeutic approaches to achieve optimal health and healing.[5](pg. 16)

The successful agreement among representatives from the 25 medical schools on a definition of their field had taken months of committee work. The eight members of the NED planning team who also were members of CAHCIM were congratulated.

But this definition was not well received by many on the call. The first issue was participatory: if integrative medicine was intended to be inclusive, then why were leaders of other disciplines not asked to collaborate on establishing the definition? The second issue was substantive and went to the heart of collaboration. Where in that definition was the evidence that a practitioner of integrative medicine must, in the words of the NED vision, be "multidisciplinary and enhance(s) competence, mutual respect and collaboration across all complementary and alternative medicine and conventional health care disciplines"?

One responsibility that ACCAHC had in the NED process was to articulate and advance shared issues relating to the cultural-medical shift toward more integrated health care. That the definition of integrative medicine included no reference to the importance of knowing how to collaborate with other types of healthcare professionals and disciplines was one issue. There were others. The word "approach," for instance, did not appropriately characterize the unique contributions of practitioners licensed in the 4000-year-old discipline of acupuncture and Oriental medicine.

ACCAHC members asked their CAHCIM colleagues to change their definition of integrative medicine. The request was well received. A conventional colleague noted that the Institute of Medicine of the National Academies, in its *Report on Complementary and Alternative Medicine in the United States*, had underscored the importance of care that is "collaborative and interdisciplinary and that respects and joins effective interventions from all sources." The CAHCIM definition was formally amended to read:

Integrative medicine is the practice of medicine that reaffirms the importance of the relationship between practitioner and patient, focuses on the whole person, is informed by evidence, and makes use of all appropriate therapeutic approaches, **healthcare professionals and disciplines** to achieve optimal health and healing [amendment in bold type].[5(pg.16)]

In short, integrative medicine requires more than integration of specific therapeutic approaches. This definition was used widely in the 2009 Institute of Medicine Summit on Integrative Medicine and the Health of the Public. Practitioners must also know how to make the best use of other healthcare professionals and disciplines to achieve optimal health and healing. Notably, the amended definition also elevated the status of all non-medical, non-CAM disciplines including nurses, physical therapists, psychologists, nutritionists, and others. A definitional platform was laid for an emerging vision of an integrated healthcare system that will require all disciplines to recognize that they must work as part of respectful, collaborative teams.

A brief coda: Several years of living with the definition have revealed one other area where some in the complementary healthcare community might prefer an additional amendment. While the definition does not say that it only refers to practice by medical doctors, many read the definition as MD-centric. Clarity that it is the "practice of medicine, *by members of any discipline*" would be a more inclusive framing.

Task Force was that the process of shifting the disciplines out of their historic isolation would benefit from a collaboratively-developed guiding document.

Through a series of phone meetings, a small work group created a draft statement of core values. These were coupled with examples of knowledge and skills associated with each of the values. The Task Force as a group endorsed the statements as a working document. The core values served as the basis of an intensely interactive World Café format for the 75 educators at the onsite NED meeting. (See the text of the draft document in Sidebar A.2.) The participants at the gathering proved to be a long way from consensus on the document. There were contradictory feelings about the ultimate contribution such a document might have, even though an onsite survey strongly supported continuing the work "to create a concise statement of core values which have resonance across the disciplines and can guide efforts to create quality integrated healthcare education."

A reconstituted Task Force continued work via a series of phone meetings after the onsite meeting. They found two resources of particular value in guiding their activity together. One was the section on practitioner to practitioner relationships in the Pew-Fetzer seminal document on relationship-centered care. (Further information on the report's findings is provided in Table A.4.) Particularly noteworthy were issues related to knowledge of self, ability to understand the "historical power inequities across the professions," "affirmation of the value of diversity," and the importance of team dynamics.[12]

The second guiding document for the team's work was the Four Quadrants from Ken Wilbur's *Integral Theory of Consciousness*.[13] The group quickly noticed that much of the integration activity has been shaped by the right quadrants, which are related to exterior inputs such as randomized controlled trials, publications, and capacity to participate. The Values Working Group engaged an internal process to explore the way that subjective issues such as self-knowledge, personal involvement in health, and appreciation of the diversity of professional experience and culture influence collaborative behavior.

Model Education Tools on the Disciplines

We recognize that expanding and strengthening integration and collaboration will require some remedial education. Practitioners of one healthcare system cannot effectively collaborate with practitioners of other systems unless they first possess accurate and adequate knowledge of and understanding about the various healthcare disciplines that exist. (This book reflects one effort to correct shortcomings.) The decades of separation between conventional and complementary healthcare disciplines, between the mind and body, and between the physical locations of our educational institutions and clinics, underscore the importance of such education. A well-formulated survey course, or set of courses, would provide an important basis for opening hearts and minds to the "other," with the ultimate and critical goal of serving patients more effectively.

There are numerous strategies to enhance collaboration and to train all practitioners about disciplines other than their own. Dialogue is vital. As practitioners become better able to discuss healthcare options outside of their own fields with both patients and other practitioners, they can better understand treatments patients may already be receiving from other healthcare practitioners. Cross-referral will be enhanced as practitioners develop openness to other medical cultures and paradigms. The potential to work successfully in multidisciplinary healthcare settings and on collaborative research projects will be enhanced.

Sidebar A.2

Toward Core Values to Shape Collaboration in Integrated Care

Note: A multidisciplinary team of educators from eight disciplines collaborated in a series of conference calls to propose the following as a set of values to guide disciplines in working toward integrated health care. While never formally endorsed by the disciplines involved in the National Education Dialogue meeting, these remain a good framework for collaboration.

Wholeness and Healing
We acknowledge and value the interconnectedness of all people and all things. We believe that healing is an innate, although sometimes mysterious, capacity of every individual, which provides a meaningful opportunity for growth and balance, even when curing may not be possible.

Clients/Patients/Families
We value our clients/patients/families as the center of our practices. We act in service to validate individuals' full experience of their wellness and illness, provide education about the possibilities for change, diminish dependence, bolster resilience, support the mobilization of their full resources, and reinforce self-reliance as important to optimal healing.

Practice as Combined Art and Science
We value competent, compassionate, relationship-centered practice that honors intuitive knowledge, stimulates creativity in the face of divergent circumstances, and appropriately utilizes a network of diverse practitioners as integral to the practice of the healing arts. This approach is informed by in-depth education, critical thinking, reliable evidence from well-designed quantitative and qualitative research, and by respect for diverse theories and world views.

Self-Care of the Practitioner

We value a practitioner's commitment to self-reflection, self-care, and to his or her personal growth and healing as being essential to one's humble ability to provide the most effective and enduring health care to others.

Interdisciplinary Collaboration and Integration

We embrace the breadth and depth of diverse healthcare systems and value collaboration of all providers within and across disciplines and with clients/patients and their families. We believe that fostering integrative and collaborative practice is essential to the creation of health, the advancement of health care, and the well-being of society.

Our Health System

We value the recognition that all practitioners have responsibility to participate in activities which contribute to the improvement of the community, the environment, and the betterment of public health. We support access for all populations to competent client/patient-centered care, which is focused on health creation and healing. We believe that advancement and promotion of health and the prevention of disease are fostered by aligning resource investment with these values.

Attitudes and Behaviors that Promote Health, Wellness, and Change

We value attitudes and behaviors which demonstrate respect for self and others and which are informed by an abiding humility, in light of both the individual care and system challenges which we face. We value authentic, open, and courageous communication. We believe that such traits are important for all participants in health care, whether practitioner, client/patient/family, educator, administrator, or policy-maker.

Source: Developed through an ACCAHC and National Education Dialogue team chaired by Carla Mariano, RN, EdD, BC-HN and including Michael Goldstein, PhD, Ben Kligler, MD, MPH, Karen Lawson, MD, David O'Bryon, JD, Mark Seem, LAc, Dawn Schmidt, LMP, Pamela Snider, ND, Don Warren, ND, DHANP, and John Weeks.

Table A.4

Practitioner to Practitioner Relationships in Relationship-Centered Care*

Area	Knowledge	Skills	Values
Self-awareness	• Knowledge of self	• Reflect on self and needs • Learn continuously	• Importance of self-awareness
Traditions of knowledge in health professions	• Healing approaches of various professions and across cultures • Historical power inequities across professions	• Derive meaning from others' work • Learn from experience in a healing community	• Affirmation and value of diversity
Building teams and communities	• Perspectives on team building from the social sciences	• Communicate effectively • Listen openly • Learn cooperatively	• Affirmation of mission • Affirmation of diversity
Working dynamics of teams and communities	• Perspectives on team dynamics from the social sciences	• Share responsibility responsibly • Collaborate with others • Work cooperatively • Resolve conflicts	• Openness to others' ideas • Humility • Mutual trust, empathy, support • Capacity for grace

*Source: Tresolini C.P. and the Pew-Fetzer Task Force. *Health Professions Education and Relationship-Centered Care.* San Francisco, CA: Pew Health Professions Commission. 1994; p. 36. ©Pew Health Commission and the Fetzer Institute.

Sidebar A.3 presents a brief overview of suggested topics and formats for educating conventional medical students about the CAM disciplines. It is based on work led by Elizabeth Goldblatt PhD, MHA/PA and Daniel Seitz, JD, involving a team of eight professionals from four disciplines; the information was presented at the NED onsite gathering. To transmit adequate knowledge in an academic setting, representatives of each discipline recommended at least 2-3 credits or 30-45 course hours on each discipline. However, already bulging educational requirements in all the health professions schools effectively vetoed such an approach. A compromise recommendation,

Sidebar A.3

Topics and Suggested Formats in Education on the Distinctly Licensed Complementary and Alternative Healthcare Professions

Educational topics

- Basic philosophy about health and healing
- Core values of the discipline
- Basic theory of the discipline and any unique terminology
- History and current status of the field
- Status of research in the field, review of the literature, evidence-based material, cultural issues that may restrict or enhance research, and research priorities identified by the profession
- Strengths and limitations in preventing and treating illness and promoting health
- Educational training requirements, competencies
- Institutional and programmatic accreditation practices
- Certification and examination agencies and processes
- Professional licensing and scope of practice
- Credentialing issues in hospitals and third-party payment
- Experience in co-management and referral strategies
- Best practices in integration with other practitioners
- Bibliography of recommended additional resources

Activities and educational formats

- Lecture and demonstrations on how the discipline approaches specific cases (best practices, strengths/weaknesses, working alone, working collaboratively)
- Joint case discussions in which specific cases are presented from the perspectives of a variety of disciplines
- Direct experience of different forms of healthcare, such as receiving acupuncture and Oriental medicine treatments, chiropractic adjustments, nutrition and diet assessments, massage therapy, and the whole person naturopathic intake
- Grand rounds presentations in integrative clinics
- Present a few simple techniques that can be incorporated into self-care
- Explore how to work collaboratively and co-manage cases with other healthcare providers

Based on materials developed by Elizabeth (Liza) Goldblatt, PhD, MPA/HA and Daniel Seitz, JD, MAT, with the ACCAHC and NED team, which also included Frank Zolli, DC, EdD, Adam Perlman, MD, Pamela Snider, ND, Don Warren, ND, and John Weeks.

though still a challenge to fit, is a survey course of 30-45 contact hours in which all of the licensed complementary healthcare disciplines are examined. Recognizing that even this is a push for many educational programs, ACCAHC hopes that this *Desk Reference* will help to increase general knowledge of the licensed complementary, alternative and integrative health professions and provide a resource for such a course.

Priority Recommendations for Interdisciplinary Educational Collaboration

After 15 months of planning and collaborative project execution, three days onsite, and two surveys of participants, the National Education Dialogue to Advance Integrated Health Care identified nine priority areas for action in which there was significant common ground (>80% agreement):[5(pg. 3)]

- Facilitate development of inter-institutional relationships and geographically-based groupings of conventional and CAM institutions and disciplines in diverse regions. Promote student and faculty exchanges, create new clinical opportunities, facilitate integrated postgraduate and residency programs, and provide opportunities for students to audit classes and share library privileges.
- Create resource modules on teaching about distinct CAM, conventional, and emerging disciplines (approved by the disciplines), which can be used in a variety of formats—from supporting materials in such areas as definitions and glossaries to full curricular modules.
- Share educational and faculty resources and information on inter-institutional relationships, including samples of existing agreements and existing educational resources through development of a website.

- Continue multidisciplinary work to create a concise statement of core values that have resonance across the disciplines and can guide efforts to create quality integrated healthcare education.
- Collaboratively develop and sponsor continuing education initiatives designed to draw participants from diverse disciplines.
- Create collaboratively developed educational resources to prepare students and practitioners to practice in integrated clinical settings.
- Develop an outline of skills and attitudes appropriate for those involved in collaborative integrated health care.
- Assist individuals with making institutional changes by offering support for leadership in change creation. Explore strategies for overcoming the challenges of prejudice, ignorance and cultural diversity.
- Explore third-party clinical sites that serve the underserved (such as community health centers) as locations for developing clinical education in integrated healthcare practices.

Collaboration following the NED meeting

Significant efforts were made to find philanthropic support that would allow the NED work groups to remain intact and begin addressing these priorities. Limited funds were raised to advance the values initiative and some interdisciplinary collaboration, but unfortunately the broader NED effort had to be suspended roughly a year later. The effort's epitaph was provided while NED was still robust by the chair of the Institute of Medicine committee that produced the 2005 report on complementary and alternative health care in the US: "What you are doing [with NED], this great collaborative work, is one of the most important things anyone can do to implement this report."[5(pg.1)]

Activity to advance the collaborative agenda has continued on many fronts. Leaders of ACCAHC chose to take what was originally a project of the IHPC and create an independent organization as a basis for ongoing interdisciplinary action. As noted in the introduction, ACCAHC was formally incorporated in 2008 and has gathered increased participation from leaders of the licensed complementary healthcare professions as well as Traditional World Medicines and emerging professions organizations which are engaging self-regulatory and regulatory efforts. (See the Introduction for more information.)

Cross-fertilization between the CAHCIM and ACCAHC has also continued. Executive teams met in 2006, and, in 2009, as noted above, roughly 80 leaders, evenly split between the two organizations, met for a half day to explore formal collaboration on various projects.

The International Congress for Educators in Complementary and Integrative Medicine (ICECIM) at Georgetown University in October 24-26, 2012 marked a culmination of this collaborative movement. This gathering, promoted by Aviad Haramati, PhD, an ACCAHC Council of Advisers member, was co-sponsored by CAHCIM and ACCAHC together with Georgetown University. The ICECIM planning team, program team, funding group, presenters and participants all exemplified the interprofessional principal of "horizontal" collaboration and mutual respect. An outpouring of over 230 submissions led to a multidisciplinary turnout of roughly 350 educators, researchers and clinicians. A special issue of the peer-reviewed *Explore* magazine, planned for late 2013, will feature content from the meeting.

The infrastructure for collaboration continues to grow. The ACCAHC Council of Advisers includes numerous leaders of CAHCIM. Content at the ACCAHC Biennial Meetings in 2011 and 2013 focus on areas of collaboration and include CAHCIM leaders. Increasingly, medical doctors can be found as members of ACCAHC Task Forces and Working Groups. In fact, in 2013 an integrative medical doctor, Molly Punzo, MD, a long-time member of the ACCAHC Clinical Working Group, was selected to lead an ACCAHC Task Force that is focused on the contributions that integrative health modalities and disciplines can have in meeting the patient experience, population health and cost-saving goals of the "Triple Aim" that is driving health system reform.

Competencies for Optimal Integration and the Opportunity in Interprofessional Education

Early on, ACCAHC leadership recognized that education in the ACCAHC disciplines had in common that they typically educated their students in discipline silos to go out to practice in silos. The emerging opportunities for collaborative, integrative practices called for new competencies.

In late 2009, in response, ACCAHC's leadership and its core five disciplines set as its central agenda creating what the organization published 11 months later as *Competencies for Optimal Practice in Integrated Environments.*[14] ACCAHC subsequently learned that a separate collaboration of councils of schools for medicine, nursing, public health, pharmacy, dentistry and osteopathy, the Interprofessional Education Collaborative (IPEC), had engaged a parallel process and published the *Competencies for Interprofessional Education and Practice.* An ACCAHC team, led by Michael Wiles, DC, MEd, MS, and massage educator Jan Schwartz, MA, led an effort to examine the potential integration of these two documents. The re-drafted ACCAHC *Competencies* adopted the four IPEC competencies in total, while maintaining two other competencies fields related to "Evidence-based Health Care and Evidence Informed Practice" and "Institutional Healthcare Culture and Practice." These were subsequently endorsed by national educational organizations from ACCAHC's core disciplines. IPEC was notified. That organization lists ACCAHC as a "Supporting Organization" for their work.

ACCAHC leaders viewed the patient-centered interprofessional education (IPE) movement as one that, on a principled basis, would be expected to be open to increased collaboration and involvement from the ACCAHC disciplines. The organization invested in this in multiple venues. ACCAHC was a sponsor of the major North American IPE conference in 2011, Collaboration Across Borders (CAB) III. The ACCAHC *Competencies* were the subject of a presentation. For the 2013, CAB IV meeting, ACCAHC was invited to name a member for the conference's Advisory Board. ACCAHC was instrumental in bringing IPE leaders and content into the 2012 ICECIM meeting in Georgetown.

Most importantly, ACCAHC's Board of Directors made a major organizational commitment in January 2012 to become a sponsor and active participant in the Institute of Medicine Global Forum on Innovation in Health Professional Education in which IPE figures prominently. ACCAHC chair Elizabeth Goldblatt, PhD, MPA/HA serves as a member of the forum with ACCAHC executive director John Weeks also active as an alternative. Goldblatt was featured in a 2012 IOM "Spotlight" and was selected as part of a seven-person team to develop content for the Forum's 2013 workshops.

Leadership in Knitting Together the Integrative Health and Conventional Disciplines

A critical, long-term ACCAHC strategy for fostering better patient care through fostering mutual respect and inclusion among the disciplines is to "show up" in major healthcare dialogues. As ACCAHC has developed capacity through growing core membership, and more importantly, through philanthropic support that accounted for roughly 80% of operating revenues in 2012, the organization has focused on participating in core dialogues. A parallel and necessary objective has been to develop a cadre of leaders who can represent integrative health values, practices and disciplines in such dialogues.

Examples are growing. For instance, via a series of letters and meetings, ACCAHC leaders played an important role in shaping aspects of the 2011-2015 strategic plan for the NIH NCCAM. Perspectives of ACCAHC participants are credited with elevating the visibility of complementary and alternative medicine in the plans for the Patient Centered Outcomes Research Institute (PCORI). ACCAHC involvement in the IPE movement and the IOM Global Forum, noted above, has helped widen the circle of the types of practitioners viewed as valuable in team care. As noted above, ACCAHC is recognized as a supporting organization for the Interprofessional Education Collaborative. The development of the Pain Action Alliance to Implement a National Strategy (PAINS) as a national, multidisciplinary campaign to promote an integrated view of pain treatment has been shaped in part through the work of massage researcher Martha Menard, PhD, CMT, ACCAHC's representative to the PAINS steering committee. ACCAHC executive director John Weeks was invited as an expert consultant to participate in two sessions of World Health Organization teams convened to develop a 2014-2023 strategic plan for traditional and complementary medicine.

The culture is increasingly open to such perspectives, even as ACCAHC is seeking to develop the capacity to take advantage of the emerging opportunities to advance the kinds of collaboration which will lead to better patient care.

Summary

The past decade has seen significant development of the values, definitional commitments, and operational infrastructure needed to support collaborative action and prepare the way for an integrated healthcare system. Conventional medicine, integrative medicine, and the licensed complementary, alternative, and integrative healthcare professions must all continue to work on that goal for many years to come.

Collaborative action is already expanding despite the historic estrangement between many of these fields. An ongoing challenge is to show how patients can benefit when practitioners of different stripes are not merely viewed as purveyors of therapies but as exponents of entire disciplines with their own rich histories and distinct approaches to patient care. Strategies for enriching cross-disciplinary experiences in the midst of already excessive curriculum requirements must be developed. To be urging additional coursework and clinical exposures on heavily burdened health professions students puts all of us in an uncomfortable position. Yet multiple forces in the broader culture energize this effort. These range from cultural shifts and consumer interests to top-level reports on the importance of collaboration in the treatment of costly chronic diseases. Patient-centered care, patient-practitioner partnerships, and team medicine are all elements of renewed efforts to create a medical system which is proactive and focused on health creation.

A frequently quoted opinion among academic researchers concerning integrative and complementary healthcare themes is that one day we will no longer have conventional medicine and alternative medicine, but only good medicine. Such a rosy resolution of the philosophical debates on diverse approaches to health and medicine that go back not just decades but centuries cannot be expected in the near future. However, we can all learn to collaborate more respectfully and effectively despite our differences, and that is certainly a part of good medicine and very likely a necessary antecedent to greater success overall.

Resources

Organizations and Websites

- Academic Consortium for Complementary and Alternative Health Care
  www.accahc.org
- Consortium of Academic Health Centers for Integrative Medicine (CAHCIM)
  www.imconsortium.org
- Oregon Collaborative for Integrative Medicine
  www.o-cim.org

Citations

1. Freshman B, Rubino L, Chassiakos YR, eds. *Collaboration Across the Disciplines in Health Care.* Sudbury, MA: Jones and Bartlett; 2010.
2. National Institutes of Health. *White House Commission on Complementary and Alternative Medicine Policy.* NIH Publication 03-5411; ISBN 0-16-051476-2; 2002.
3. Institute of Medicine of the National Academies. Committee on the Use of Complementary and Alternative Medicine by the American Public, Board on Health Promotion and Disease Prevention. *Complementary and Alternative Medicine in the United States.* National Academies Press, 2005.
4. Quinn, S, Traub M, eds. National Policy Dialogue to Advance Integrated Health Care; Finding Common Ground. Integrated Health Care Consortium. Integrated Health Care Consortium; 2002. http://ihpc.info/resources/NPDFR.pdf; accessed August 24, 2008.
5. Weeks J, Snider P, Quinn S, et al. *National Education Dialogue to Advance Integrated Health Care: Creating Common Ground.*

Progress Report, March 2004-September 2005. National Education Dialogue Planning Committee, Integrated Healthcare Policy Consortium; (2005). http://ihpc.info/resources/NEDPR.pdf; accessed August 24, 2008.

6. Eisenberg DM, Kessler RC, Foster C, et al. Unconventional medicine in the United States -- Prevalence, costs, and patterns of use. *NEJM* 1993: 328:246-252.

7. Eisenberg DM, Davis RB, Ettner SL, et al. Trends in alternative medicine use in the United States, 1990-1997: results of a follow-up national survey. *JAMA* 1998;280:1569-1575.

8. Nedrow, AR, Heitkemper M, Frenkel, M, et al. Collaborations between allopathic and complementary and alternative medicine health professionals: four initiatives. *Acad Med.* 2007; 82: 962-966.

9. Weeks J. The future of MD education: empathy, roses and respect via Georgetown and U Minnesota pilots. *The Integrator Blog News & Reports*. August 15, 2006. http://theintegratorblog.com/site/index.php?option=com_content&task=view&id=128&Itemid=93; accessed August 24, 2008.

10. Kligler B, Maizes V, Schachter S, et al. Core competencies in integrative medicine for medical school curricula: a proposal. *Acad Med.* 2004; 79:521-531.

11. Benjamin PJ, Phillips R, Warren D, Salveson C, et al. Response to a proposal for an integrative medicine curriculum. *JACM.* 2007:13:1021-1033.

12. Tresolini, CP and the Pew-Fetzer Task Force. Health Professions Education and Relationship-centered Care. San Francisco, CA: Pew Health Professions Commission; 1994:36. http://www.futurehealth.ucsf.edu/pdf_files/RelationshipCentered.pdf; accessed August 24, 2008

13. Wilbur K. An integral theory of consciousness. *Journal of Consciousness Studies*. 1997; 4:71-92.
http://www.imprint.co.uk/Wilber.htm; accessed August 24, 2008

14. Academic Consortium for Complementary and Alternative Health Care (ACCAHC). Competencies for Optimal Practice in Integrated Environments. 2011.
http://accahc.org/images/stories/accahccompetencies_120612

Appendix 4

ACCAHCcronyms
Academic Consortium for Complementary and Alternative Health Care Acronyms

Councils of Colleges
AANMC - Association of Accredited Naturopathic Medical Colleges
ACC- Association of Chiropractic Colleges
AFMTE – Alliance for Massage Therapy Education
AME – Association of Midwifery Educators
CCAOM - Council of Colleges of Acupuncture and Oriental Medicine

Accrediting Agencies
ACAOM – Accreditation Commission for Acupuncture and Oriental Medicine
CCE - Council on Chiropractic Education
CNME - Council on Naturopathic Medical Education
COMTA - Commission on Massage Therapy Accreditation
MEAC - Midwifery Education Accreditation Council

Testing and Certification Organizations
NABNE - North American Board of Naturopathic Examiners
NARM – North American Registry of Midwives
NBCE – National Board of Chiropractic Examiners
NCBTMB - National Certification Board for Therapeutic Massage and Bodywork
NCCAOM - National Certification Commission for Acupuncture and Oriental Medicine

Emerging and Traditional Professions

ACHENA - Accreditation Commission for Homeopathic Education in North America
CHC - Council for Homeopathic Certification
IAYT - International Association of Yoga Therapists
NAMA - National Ayurvedic Medical Association
YA - Yoga Alliance

Disciplines

AOM – Acupuncture and Oriental Medicine
DC – Chiropractic
DEM – direct-entry midwifery
EP – Emerging Profession
IM – Integrative Medicine
MT - Massage Therapy
ND – Naturopathic Medicine
TWM – Traditional World Medicine

Other Acronyms

Within ACCAHC

AA SPIG - Accreditation Agency SPecial Interest Group
CC SPIG – Councils of Colleges SPecial Interest Group
CEDR – Clinicians and Educators Desk Reference
CT SPIG – Certification/Testing SPecial Interest Group
EC – Executive Committee
HSCP – Hotspots/Cooling Point
SPIG – SPecial Interest Group

Outside ACCAHC

ABIHM - American Board of Integrative Holistic Medicine
ABOIM - American Board of Integrative Medicine
AHMA – American Holistic Medical Association
AHNA - American Holistic Nurses Association
AMTA - American Massage Therapy Association

CAHCIM – Consortium of Academic Health Centers for
Integrative Medicine
ICC-CIM International Congress for Clinicians in
Complementary and Integrative Medicine
ICE-CIM International Congress of Educators in Complementary
and Integrative Medicine
IHPC – Integrated Healthcare Policy Consortium
NCCAM - [National Institutes of Health] National Center for
Complementary and Alternative Medicine
NED – National Education Dialogue to Advance Integrated
Health Care

For information about ordering additional copies, please visit:

www.accahc.org